D0629809

HOPE
ON EARTH
A CONVERSATION

HOPE
ON EARTH
A CONVERSATION

PAUL R. EHRLICH & MICHAEL CHARLES TOBIAS

with additional comments by John Harte

The University of Chicago Press
Chicago and London

PAUL R. EHRLICH is the Bing Professor of Population Studies and the president of the Center for Conservation Biology at Stanford University. He is the author or coauthor of many books, including *The Population Bomb*; *The Dominant Animal: Human Evolution and the Environment*; and *Humanity on a Tightrope: Thoughts on Empathy, Family, and Big Changes for a Viable Future.*
MICHAEL CHARLES TOBIAS is an ecologist, author, filmmaker, and president of the Dancing Star Foundation, a nonprofit organization based in California and focused on international biodiversity conservation, global environmental education, and animal protection. His works include *World War III: Population and the Biosphere at the End of the Millennium*; *Sanctuary: Global Oases of Innocence*; and the recent feature-film trilogy *No Vacancy*, *Mad Cowboy*, and *Hotspots.*

The University of Chicago Press, Chicago 60637
The University of Chicago Press, Ltd., London
© 2014 by The University of Chicago
All rights reserved. Published 2014.
Printed in the United States of America

23 22 21 20 19 18 17 16 15 14   1 2 3 4 5

ISBN-13: 978-0-226-11368-5 (cloth)
ISBN-13: 978-0-226-11371-5 (e-book)
DOI: 10.7208/chicago/9780226113715.001.0001

Library of Congress Cataloging-in-Publication Data

Hope on Earth: a conversation / Paul R. Ehrlich & Michael Charles Tobias; with additional comments by John Harte.
      pages; cm
   Includes bibliographical references.
   ISBN 978-0-226-11368-5 (cloth: alk. paper) — ISBN 978-0-226-11371-5 (e-book)
1. Ecology. 2. Environmental ethics. 3. Global environmental change. 4. Nature conservation. I. Ehrlich, Paul R. II. Tobias, Michael. III. Harte, John, 1939–
   QH541.145.H665 2014
   577—dc23   2013035831

♾ This paper meets the requirements of ANSI/NISO Z39.48-1992 (Permanence of Paper).

CONTENTS

*A gallery of photographs
follows page 32.*

PRELUDE

Amid melting snowfields, a profusion of colorful butterflies and other pollinators, and a riotous display of wildflowers, native and non-native to the Rockies, Paul R. Ehrlich and Michael Charles Tobias met for a couple of days at the Rocky Mountain Biological Laboratory in the mountains above Crested Butte, Colorado, to hike together and discuss the fate of the world. They focused on the ethical ambiguities and underpinnings whose topical urgency could generate many encyclopedias of data and analyses, and discussed some of the research being done to further understanding of biological

Gothic, Colorado, site of Rocky Mountain Biological Laboratory. © M. C. Tobias

dimensions of the human predicament. However, this modest volume, *Hope on Earth*, is intended to be a reflection only upon those points of view, and conceptual, scientific, and ethical opinions that were first discussed on those mountain hikes, and which continued for the year to follow.

The end result—growing out of a year of subsequent additions to the book, reflecting, in part, world events—is presented here. Given the dynamism of all of these topics, this book is but a snapshot—an incomplete document of feelings, opinions, and priorities. It is our joint attempt to add nuance, candor, and personality to topics that all too often are limited to the podium and other conventional public forums or, worse yet, not included in public discourse. We hope through this format that we will inspire you, our readers, to pick up threads of the conversation and carry them far and wide.

HUMAN HEGEMONY: THE ONE VERSUS THE MANY

Shoppers in Shanghai. © M. C. Tobias

MICHAEL TOBIAS, *hereafter* **MT:** Looking up at those cliffs and talus slopes, waterfalls and snowfields, with the fair breeze of a high-altitude Colorado summery morning, with the birds singing and our local world carpeted by gorgeous flowers and adrift with attendant butterflies, there would hardly seem to be any justification for griping. We're in paradise at nearly 10,000 feet.

But we are also in global hell, up against some incredibly high stakes—nothing short of the fate of the Earth. What are the core issues of the conscience, activism, and idealism that underlie—as well as those undoing, undermining, and confusing—the debate about animal rights, biological conservation, and the stakes for the future of life on Earth? That's what is thoroughly nagging at my scientific and intuitive clockwork.

Pursuing that direction, let me say that one of the underlying premises that I hear constantly from committed individuals of the conservation biology world and of the animal rights, animal protection, and animal welfare worlds (all of whom are in some form of slight or radical disagreement over levels of protection) is as follows: If you focus in your own personal life upon saving individuals, that's going to be about as much as you can do. Applying your knowledge, experience, skills, data sets, and compassion to an individual who needs you is all-encompassing, typically exhausting, and, yes, usually deeply rewarding. But it is not going to provide you much energy or time for field research that would give you the information needed to help save populations or even whole species. It's not that saving an individual is fundamentally different than saving an entire species—but, in truth, it is, and we all know it.

It is an ancient Greek paradox that Plato elaborated upon in his dialogue *Parmenides*, which weighed in on the great Eleatic meeting of Parmenides and Zeno and their debate regarding this "one versus the many." Of course, humanity has grappled with the dialectic ever since. We're still in the throes of it, by all appearances.

PAUL EHRLICH, *hereafter* **PE:** This reminds me of the efforts made on behalf of the oiled birds during the BP Deepwater Horizon Gulf disaster. You just could not save them all. And it is an even more difficult problem when the issue is which of many species to save, which is increasingly the case. And this obvious but intractable dilemma is not unlike the paramedics' paradox of World War I that poet e. e. cummings experienced as a volunteer ambulance driver on the front: Whom do you save? How do you determine and justify the candidates for triage? And that dilemma becomes even more horrible when one considers that today we are already triaging human populations (who gets fancy cars and clean water versus whose children must walk far to gather firewood and could die of waterborne diseases), and that situation is likely to get *much* worse.

MT: Precisely. I viewed the HBO documentary *Saving Pelican 895*, which shows efforts to save what avifauna could be spared, but at least 7,000 birds died following that April 2010 British Petroleum oil spill. I'm sure that the numbers are going to escalate as more and more biological opinions come in over the years. The 2012 book by Antonia Juhasz, *Black Tide: The Devastating Impact of the Gulf Oil Spill*,[1] chronicles, under the Freedom of Information Act, countless health problems, both for wildlife and humans, in particular assessing reports from the Joint Unified Command Center in Houma, Louisiana, and those of the U.S. Coast Guard.

In the case of the *Exxon Valdez* disaster on March 24, 1989, the biological fallout continues. It's hard for me to absorb

1. Antonia Juhasz, *Black Tide: The Devastating Impact of the Gulf Oil Spill* (New York: John Wiley & Sons, 2012). *Black Tide* is also the title of the Discovery Channel film on the Exxon disaster; I directed the film some six months after the terrible event in question.

what transpired, even though I made a film about that crisis. I waded ankle-deep in oily mousse six months after the accident, and 1,500 miles to the west of the *Valdez* spill in the Aleutian Islands, I saw yet more dead oiled birds. Today's undergraduate students weren't even alive at that time. Yet we do not seem to learn from our mistakes as a culture. We pass down these terrible environmental legacies, but the meaning—the substance—eludes us. Director Rob Cornellier followed up in 2009 with his own documentary, *Black Wave: The Legacy of the Exxon Valdez*, and revealed how twenty years later citizens of Cordova, Alaska, were still dealing with the ecological impacts. These things stay with us, even if the majority of humans are far removed. Which speaks again to the issue of individuals versus groups; saving individual birds versus protecting their entire colony, rookery, habitat, or elements of the global avifauna that depend on the area during migration.

In March 2012 the French firm of Total SA was described by the *Wall Street Journal* as having "reported a sheen of gas condensate six nautical miles wide around the Elgin platform. The leak is thought to be originating below the pumping platform at the well bore-head. The gas is flowing up the piping structure onto the platform, spewing out as gas, with some turning to liquid," about 150 miles off the coast of Scotland's city of Aberdeen.[2]

The peril to wildlife will only escalate as humanity's hegemony escalates. Add yet another component, that of aging pipelines. In the wake of the recent March 2013 Mayflower, Arkansas, ExxonMobil spill, there are other legal complica-

2. Inti Landauro, Sarah Kent, and Alexis Flynn, "Total Plays Down Risk of Gas-Leak Explosion," *Wall Street Journal*, March 29, 2012, http://online.wsj.com/article/SB4000142405270230 340470457730919381097157o.html.

tions, such as the differing classification for tax purposes of tar sands oil versus conventional oil. The former, which spilled in Arkansas, evidently is exempt from the Oil Spill Liability Trust. This might seem like an arcane detail, but it underscores the complex wrangling of major multinationals, the IRS, and the continuing legacy of liability, particularly as concerns wildlife across the planet; wildlife that just gets more and more hammered.

PE: First of all, efforts to "rescue" wildlife probably don't work most of the time. It takes a relatively gigantic effort to save a single oiled seabird or seal. Similarly, on land the efforts to rehabilitate orphaned or injured great apes are fraught with all kinds of difficulties. For any rehabilitated animals, the issue of successful reintroduction to the wild is ordinarily problematic. Especially with great apes, our own gang, the ethical issues of what's worth the effort are especially heartrending. One major question is, how much do you count the opportunity costs: Would the same effort do more for Earth's flora and fauna, *and for us and the other great apes*, if directed elsewhere?

OPPORTUNITY COSTS

MT: How would you define "opportunity cost"?

PE: Opportunity cost is what you've forgone to make the effort; what you think would be the second best use for funds versus the one you've chosen. If you spend money rehabilitating injured apes, the opportunity cost would be that money not being available to protect ape habitat, which you consider the best alternative.

MT: So, the cost of an alternative that needs to be scrapped in order to pursue some other course of action, right?

PE: Yes, and, of course, what is the second best is always another issue. If instead of putting your time into saving seabirds one at a time, you spent your equivalent time painting rooftops white or deploying solar technology or lobbying Congress to get rid of offshore drilling (since we shouldn't burn the oil anyway), would you advance what are ultimately your goals better? Those are the sorts of issues that need to be considered when deciding to put the effort into saving one individual at a time.

THE ORIGINS OF ANIMAL RIGHTS: WHY A PELICAN AND A CHICKEN MATTER

MT: It's the "one individual at a time" reality that, in my opinion, is the veritable core of all animal rights.

PE: Explain?

MT: You connect with an individual; you empathize not with an alien statistical population, but with a face, a heartbeat, an animal that you are, perhaps, holding in your arms, whether an oiled pelican or a rescued chicken.

PE: Why chickens?

MT: I mention chickens because they are the most numerous of animals slaughtered for human consumption.

PE: Insects hitting car windshields don't fare too well.

A rooster and human friend's finger. © M. C. Tobias

MT: From a Jain ethical perspective, yes, the insects are clobbered every day. I don't dispute insect sensibilities. But with chickens, I believe it is fair to declare that hundreds of millions of people deliberately collaborate in their horrible destruction; birds no different than birds we commemorate, like the rare parrot groups or eagles. With chickens, ten billion per year, just in the United States, are subjected to horrors comparable to Auschwitz.

U.S. corporations profited to the tune of $134 billion for the year 2011 from the killing of chickens, in addition to many other vertebrates, like cows, pigs, sheep, and turkeys.

Yet, like pelicans or bald eagles or rare parrots, they are birds, with intelligence, feelings, and long-term memory—dozens of well-documented specifics and hundreds of important peer-reviewed footnotes in the scientific literature all commending the myriad of cognitive superlatives in chickens (*Gallus gallus*) and all birds.

PE: On the one hand, there are a lot of things you can do in enhancing animal rights that are not basically the saving of a creature one at a time. For instance, you could push for laws that declare, "You can't beak-clip chickens."

MT: Well, yes. For starters.

PE: Right. But, on the other hand, the difficulties mount if you want regulations that mandate there be at least six square feet of space outside for each individual bird versus one three-yard-by-three-yard open area for ten thousand "free-range" chickens. If, as is normal, the chicken feed is supplied inside and they're not going to go outside anyway, is there anything ethical about the "free range"?

MT: I have countless issues with that. My book *God's Country: The New Zealand Factor*, written with my wife, Jane Gray Morrison, is a 600-page analysis of such laws, especially with chickens. We examine the same "hubris" that features as the title of a recent documentary on the U.S. war in Iraq that showed repeatedly on MSNBC. Our contention is that it is human hubris that figures more than anything in the assumptions underlying the notion of "free range." That is a term inappropriately used by those who seek, ultimately, to kill and eat the chicken, justifying their consumption of the innocent animal by suggesting that free range is somehow natural. But Jane and I are firm in our belief that this is simply a justification by the consumer; a false sense of feeling ethically consoled when, in truth, an animal will be murdered and devoured whilst someone makes a profit on the killing and the consumer feels they are environmentally sensitive to the needs of a chicken in nature. This is just insane, and only those in the animal rights movement actually seem to understand.

PE: Well, let me just say that chickens aren't endangered. They are the most abundant, widely distributed bird on the planet.

MT: So what if they are?

PE: That's relevant.

MT: Is it?

PE: If you are concerned about optimizing those you can save versus those who are less imperiled planet-wide, yes.

MT: Paul, I think those are not the most relevant qualifications given the cognitive, neurological, and hence emotional complexities of these birds. Humans are not endangered, either. But that would certainly not diminish the consequences, for example, of an all-out nuclear exchange between North Korea and the United States.

Taking it to another level, John James Audubon also firmly believed early on in his career that the passenger pigeon, now extinct, was doing just fine. And that raises an interesting basis for discussion in terms of not just opportunity costs but prioritizing ethical imperatives. Corporate culture never treats opportunity cost analysis in financial statements because, obviously, there is no palpable value there.

PE: Of course there is. All sorts of values: the value, for example, of ecotourism to a high-biodiversity asset destination—the Amazon, as viewed by tour operators, versus the South Side of Chicago.

MT: But when one speaks of the next best purpose, we are quite clearly in the realm of playing God, certainly when discussing deliberations about life and death, and strategies for optimizing whatever time, human resources, dollars and cents, tactical supplies, and wherewithal that can be brought to bear on any given situation. And then there is what I would call the lost ethical opportunity cost—the opportunity for humans and chickens to coexist more effectively in terms of something to be gained other than a chicken sandwich.

Indeed, if ever there were a vacuum created by humans, it is the chicken sandwich, and the vacuum in nature that has resulted from the pointless, cruel, and totally unnecessary demise and suffering of that chicken, of which there are some twenty-five billion of various breeds at any one time in the world. Yes, as you rightly point out, they are abundant, but that abundance has been their tragedy: the most intensely concentrated pain meted out to creatures that can possibly be envisioned. Their biological success is their very downfall.

PE: That could be said of humans, as well. Any sober environmental scientist knows that is likely to be true.

MT: I concur. Obviously. But then, we've got *so many ethical wildfires* in the world to put out, that it's often the case that one individual activist can all too readily simply shut down from battle weariness, from what former United Nations secretary-general Boutros Boutros-Ghali once termed "compassion fatigue": that inability of organizations like the United Nations to agree on anything, let alone get into the field and make meaningful advances. He was, of course, being modest. And the latest document that came out of the Rio+20 Summit, *Realizing the Future We Want for All,* can be interpreted for better, or for worse. I'll go with better.

But it still does not obviate the truth of ten billion slaughtered chickens in the United States annually. That represents a colossal gap in any ethics-based system that purports to be having effective representation across the entire landscape of sentient beings.

PE: Each of us picks our battles, does our best. Many do nothing, others worse. But we're only human beings, and human natures are imperfect, that is a certainty. Anyway, we probably have a deeper disagreement here. I always keep in mind that ethics themselves are *entirely* a human invention, and how anthropocentric they are in comparison to ecocentric is simply a result of cultural evolution at a population level and can also be a developmental process within individuals. Attitudes toward animal abuse have changed dramatically in the recent past. Two hundred years ago, you could beat your horse to death if it didn't pull your carriage as fast as you wished—our society doesn't permit such behavior now (indeed, in the pre–Civil War South, you could beat one of your slaves to death, since he/she was your property).

Animal rights seem likely to be much expanded in the future, especially as the health, ecological, and hunger consequences for humanity of increasing the use of meat become clear. But in my own development, while I have reduced my red meat consumption dramatically, I haven't transitioned to vegetarianism, even though at times I think it would be more ethical. My view of chickens changed somewhat getting to know some kept as pets and egg-layers by friends, but I'm still not bothered by eating a chicken sandwich or a poached egg. And I'm unimpressed by the claim that it's unethical to eat eggs or even to use animal feces as fertilizer.

Inconsistent, yes, but we all struggle with inconsistency all the time. I was recently enraged by the news that state legislators have been moving to make taping of cruelty to animals on farms criminal behavior. This is perfectly understandable when you realize that in the United States legislators are largely owned by industries. (Murder Incorporated units such as the arms and cigarette industries are outstanding examples.) What's less understandable is the legislators' current crusades against women—which in my ethical view are much, much worse than their disgusting protection of animal cruelty. But both underline the human invention of ethics—one person's ethical act (forbidding women to have an abortion) is another person's (my) highly unethical act. But there are so many fronts that require ethical action. . . .

MT: I think we can multitask effectively, if we choose to. I can be a vegetarian *and* donate to charities *and* help somebody across the street, et cetera.

PE: And I can enjoy a chicken sandwich *and* work, as I have my entire life, to make this, however incrementally, a better world. Considering the vast variety of human cultures and the even wider variety of personal histories, it seems inevitable (and likely permanent) that people will invent and adopt somewhat different sets of ethics. The key seems to me to determine important common ground and keep the evolution moving toward a sustainable and happier world. In that world I would care more about the maintenance of biological diversity, and you more than I on the humane treatment of individuals of the species we domesticate or otherwise interact with—even though I still care about that. I don't like killing butterflies in the course of my research, but it doesn't bother me so much that I won't do it when necessary; and I've no qualms about stepping on a roach or slapping a mosquito, even though there are those who do.

MT: This is where those "Ask Anything" websites come in handy. For example, the so-called "Decision Innovation" website,[3] which is chock-full of relevant famous quotations about making decisions and learning from the past. Consider: "We've all heard that we have to learn from our mistakes, but I think it's more important to learn from successes. If you learn only from your mistakes, you are inclined to learn only errors" (Norman Vincent Peale); and "He who joyfully marches in rank and file has already earned my contempt. He has been given a large brain by mistake, since for him the spinal cord would suffice" (Albert Einstein).

PE: And George Santayana paraphrasing Edmund Burke to the effect that those of us who ignore the lessons of history are doomed to repeat them.

MT: Or in reincarnation-oriented cultures, those who kill a chicken may well be reborn as that chicken just before it is killed. But in terms of multitasking on positive ethical fronts, I was looking at a *Forbes* interview of UN Foundation CEO Kathy Calvin by Rahim Kanani. Ted Turner had created the foundation with his $1 billion gift to the UN in 1998, and the foundation has focused on such things as the empowerment of women, clean energy, curbing child mortality, and the like. As Calvin points out in the interview, "The United Nations has made a strategic effort to put women at the top of the global agenda. If you invest in women's health, as Secretary-General Ban Ki-moon has done through his Every Woman Every Child global strategy, you are investing in the health of a family and community. This is something that many organizations

3. http://www.decision-making-solutions.com/quotes_about_mistakes.html.

around the world are beginning to realize and implement, but the UN has been at the forefront."[4]

PE: Your point?

MT: Human beings are not just one thing. Any more than our organizations and institutions need be one thing. The UN does thousands of wonderful things simultaneously. So does any biological organism, from within. At the first Rio Summit in 1992, the proportion of women and NGOs versus men and governments was relatively minor. This time, twenty years later, women and NGOs—and the hundreds of side events, which also encompassed presidents and prime ministers—were the real stars. At this Rio+20 Summit in June 2012—which I had the privilege of attending in an official speaker capacity, on behalf of Yasuní National Park in Ecuador, northwestern Amazonia—the UN again showed its true strength, in my opinion, by providing the narrative for nearly 50,000 representatives of various stripes and colors, hopes and dreams, to engage in most courteous discourse that lent, I think, a true window on the possibilities for humanity. A month later in London, at the Family Planning Summit convened by the UK government and the Bill and Melinda Gates Foundation, resources were mobilized to assist an estimated 120 million people without access to family planning around the world, a gap that had only grown since the time of the UN International Conference on Population and Development at Cairo. Another 200 million people who want contraceptive services

---

4. Rahim Kanani, "An In-Depth Interview on International Development with Kathy Calvin, CEO of the UN Foundation," *Forbes.com*, October 20, 2011. http://www.forbes.com/sites /rahimkanani/2011/10/20/an-in-depth-interview-on-international-development-with-kathy -calvin-ceo-of-the-un-foundation/.

but can't get them are now above the radar screen. So this is all good news. Add that to such other cited examples as UNICEF and the UN Refugee Agency (UNHCR), the UN Population Fund, and the World Health Organization (WHO), and one is speaking of helping people in nearly half the countries of the world each year. You've got these so-called "large-scale, multi-stakeholder advocacy" alliances in step with the Millennium Development targets for universal reproductive health, and so forth.

PE: First it was chickens, then reincarnation cultures, and now Millennium Development targets.

MT: Precisely. We can go from chickens to the International Conference on Population and Development without skipping a beat. This is part of the challenge: the proliferation of issues.

But it also is what we are capable of—whether as lunatics or poets.

THE COMPLEX CONNECTIONS

PE: Or both. It's more than merely too many issues confronting our species all at the same time. It's the problem of actually getting people to agree on the big problems, understanding their complex and myriad interconnections, and then doing something about them.

MT: We get there all the time. In every "small step . . ."

PE: Personally, I see nothing unethical about eating a chicken sandwich, but I see a lot of unethical practices in how chickens

Los Angeles 405 freeway traffic. © M. C. Tobias

are treated en masse, and how the food system in general is run, and how hundreds of millions of people on the planet can't afford to eat a chicken sandwich—who would never be faced with the ethical choice we disagree on. And that's a problem that is likely to get much worse as the population continues to expand and the agricultural system runs down in the face of climate disruption, soil deterioration, groundwater exhaustion, inadequate support of research in crop genetics and conservation of crop relatives, agribusiness greed, and myriad other problems.

MT: I'd just keep chickens out of the disagreement. Why? Because it's like saying pro-life versus pro-choice. We're losing the core in the semantics, I'm afraid. What if neo-Nazis were debating how to best frame their favorite portrait of Hitler? What if—with attendance down to statistically zero percent in church—Irish clergy were still pretending they had 100 percent attendance as a guide for their ethical decisions; or what if gun-control lobbies declared victory on the same issue the NRA claimed victory? You see, the blurred ethical summations may have little to do with the victims themselves, in this case chickens or women or the cherished memories of all the too-numerous Holocaust victims.

PE: We have to agree to disagree. That is the whole point of allegedly cultivated discourse.

MT: Correct.

PE: So we are, by definition if not outright inclination, committed to finding a thread that laces all of this chaos together meaningfully. Give me a substantive reason to even begin to focus on a solution to anything, if, as you've intimated, there are underlying impossibilities.

MT: I didn't mean to suggest there are.

PE: Sure. With a chicken becoming the gold standard for all ethical deliberations or universal imperatives . . .

MT: It is a species that is implicated, like so many other species, because of our own shortcomings, as humans in general. Consider the more than one billion hungry, malnourished humans that the World Health Organization and many others are tracking and trying to help. You'd think it would be a no-brainer: if we can proverbially land a person on the moon, we should be able to curb human hunger.

PE: I agree, but for a wide variety of both genetic and cultural reasons, our species finds it less difficult to go to the moon.

MT: But I don't think human ethics are the driving force in nature that engenders these terrible chronic malnutrition statistics or outbreaks of famine, short of deliberately inflicted Holocaust situations. I've been among tens of thousands of starving victims, in 1975 in Bangladesh, where I joined many others in handing out survival packages to long lines of dying individuals; biscuits and milk powder provided by international emergency agencies. It is a horror indelibly fixed in my mind. You never get over it, and I keep thinking: There is, of course, absolutely no excuse for hunger and starvation. We can speculate endlessly about alleged human-induced complexities, political revolutions, civil wars, declining water tables, inequities in distribution of foodstuffs, corporate monopolies and intellectual property rights, fluctuating market economies for grain, and the global truth of climate change and its effects. But again, it's not human starvation versus chickens.

PE: Obviously, there are numerous complexities. The minute we try to reduce them to a simple equation or explanation, there are bound to be odd glitches in logic.

MT: Exactly. Those odd glitches . . . so many underlying issues that have cascaded as a result of the human population explosion.

PE: You're telling me? The quintessential question that Malthus grappled with and many others.

MT: Yourself included (*smiling*).

PE: For sure. All with respect to our species exceeding its ecological carrying capacity and, as environmental scientists know, destroying its life-support systems.

MT: Agreed. Still, it is anything but logical to equate the solution to human excess with the consumption of one animal or another. I can't envision a chicken as that point of salvation between starving to death or not starving to death. But then, you're right, I am fortunate that I've never been confronted with that precise dilemma.

PE: And I think you are too easily impressed by sadly inadequate good news.

MT: What do you mean?

PE: A common example is the increase in the use of solar energy–mobilizing technologies in the face of massive and globally increasing greenhouse gas emissions from a highly subsidized and criminally uncaring fossil fuel industry and population growth expanding the number of emitters. One could, of course, draw the conclusion that a basic flaw lies in a capitalist economic system that emphasizes perpetual growth and profit over compassion.

MT: Well, it's true: I'll take anything that has a positive over a negative sign. I'm even happy if there's a little fresh water to drink. For me, other than any decline in bad news, that's better than most days. So what are you really suggesting, for example?

PE: OK. For example, while the Gates Foundation putting money into family planning is, indeed, good news, just the news of the continuing increase in the greenhouse gas concentration in the atmosphere overwhelms it. After all, we're destroying the benign climate in which agriculture and civilization developed. And that's without adding in the escalating loss of critical elements of humanity's natural capital; the accelerating toxification of the planet; what Michael Klare describes as the "Race for What's Left" of resources; the increasing danger of a civilization-destroying resource war (perhaps between India and Pakistan over Himalayan water); or the novel epidemics made more likely by increased population size combined with rapid transport systems, misuse of antibiotics, and malnutrition making people immune-compromised.

EDWARD CURTIS AND OTHER ECOLOGICAL ECHOES

MT: So you've just actually added fuel to a contradiction. This litany of bad news in no way implicates a chicken, but rather a particular dictator in Bangladesh in the mid-1970s and the cumulative consequences of human disingenuous greed, manipulation of natural resources, and any number of human-induced conflicts that add up, backfire, undermine so much of the goodness that can also be discerned in human natures, like a Gates Foundation.

As the planet heaves, human ethics will undoubtedly disintegrate. Things will get worse, in my view. Frankly, not just hunger and famine, but the billions of chickens slaughtered will also yield a telling portrait of just how terrible times have gotten.

You might think that the leap to a chicken is somehow discordant, a rude juxtaposition. But, again, I would argue: The chicken is iconic. He/she represents all that human civilizations have done to subvert, dominate, and, to a large extent, destroy the natural world and all those precious beings within. As Edward Curtis, in his twenty-volume *The North American Indian* (1907–30), sought to show in his masterpiece image, *Vanishing Race*, the assimilation and/or genocide of an entire people—the collective heart and soul and reality of indigenous North Americans.

Similarly, the chicken—all the breeds, beauty, colors, shapes, and sizes; and likely cultural-psychological-emotional variety combined—is a vanishing race. The species may not be endangered, but each individual chicken is endangered.

The literature regarding chickens is growing markedly. Twenty distinctive expressions have been identified; ten thousand years of behavioral characteristics noted. A 2005 Compassion in World Farming conference reported on a number of scientific disciplines speaking to the sophisticated "brain chemistry and neurobiology of farm animals," with one animal welfare professor at Bristol University, Dr. Christine Nicol, acknowledging that "chickens may have to be treated as individuals."

Meanwhile, the scientific literature concerning feelings and cognition in animals is immense. So many scholars continue to ignore what has become overwhelmingly of importance to true science and research, namely, the fact of consciousness and self-awareness, even ethical dispositions, in other species, domestic and wild.

So imagine the escalating contradictions: more and more chickens being slaughtered to feed a species (humans), more and more of whom are going hungry, and many of whom—at least in the United States and in specific states like Mississippi—are obese. Why, because calories are cheap. Nutrition is expensive. That's truly a preposterous situation. Yet that's what's already happening.

PE: I know. I don't disagree, but there are also millions of individuals of interesting, complex birds and mammals being slaughtered by our pet cats every day—in at least one instance even wiping out an entire avian species. At least the species *Gallus gallus* (the chicken and the ancestral red junglefowl) is relatively safe from extinction because we domesticated it— it certainly wouldn't be safe if we hadn't.

MT: But then there's this other problem, namely, trying to be positive amid so much bad news. In your book with Robert E. Ornstein, *Humanity on a Tightrope: Thoughts on Empathy, Family, and Big Changes for a Viable Future* (2011),[5] as well as your much earlier *Human Natures: Genes, Cultures, and the Human Prospect* (2000), you are not by any means entirely pessimistic. Indeed, your MAHB organization [Millennium Alliance for Humanity and the Biosphere] is thoroughly driven, as I perceive it, by the very real possibility that humanity could engender a very positive future for itself and other species.

PE: With a lot of luck and hard work. But sadly there is little sign of either. Right now, for example, the United States is busy *increasing* its use of fossil fuels with the crazy natural gas "boom," and most of its citizens seem to believe that population growth is the key to prosperity.

MT: That's what we're capable of. That's why I think the great legacy of Edward Sheriff Curtis (1868–1952) is so relevant. His life reads as one heroic, Nobel Prize–quality expedition to save the best in ourselves. Artist, explorer, anthropologist, photographic inventor, environmentalist, filmmaker . . . the man who sought to preserve the memory of the best of North American indigenous life and ecosystems.

He was, for all of his utterly forlorn situations, an opti-

---

5. For interview with Paul Ehrlich about his book, see https://www.youtube.com/watch?v=XGoG3fD7_GQ.

mist. Imagine, amid the eighty-odd Native American tribes that were fast disappearing, he believed an honest record of their biographies, their way of life, varied music, languages, diet, architecture, artwork, could rekindle a renaissance in the human spirit. Probably no other individual of the twentieth century was so obsessed with saving that which we were losing, and documenting it—in Curtis's case, some forty thousand images. The Rachel Carson of ethnography; the Paul Ehrlich of butterflies . . .

PE: I would wager that butterflies, by and large, have fared better than First Peoples and Native Americans—indeed, than virtually all of the groups overrun by the expansion of "civilizations." We're all very much aware of the tragic American genocide, less so of others. I would like to think we've learned something as a result. I'm not sure we are any closer, however, to eradicating those elements in our genetic-developmental background, which seem to predispose one civilization after another to exterminate others. There are all sorts of substitution variables. We kill one another in tens of thousands of traffic accidents, through manufacturing cigarettes, or through overdoses or civil war or with gun violence, spousal and child abuse, you name it. And, perhaps, if you look at the planet from the perspective of some terracentric omniscient demographer, the Earth welcomes pandemics, massive influenzas. It's controversial to go there. The minute we invoke human ethics, we find ourselves potentially at odds with all of the planet's critical biological watersheds. Curtis enlarged the frame of reference for grasping the beauty and importance of American indigenous peoples and their fascinating cultures and sometimes powerful connection to the land. But I wonder to what he extent he was part of the solution in terms of saving them.

MT: Well, remember, by the time Curtis launched his ambitious effort in 1906, following funding from J. P. Morgan, the Long March and so much of the tragic history had already happened. He could not vindicate genocide, only record the remaining faces, traditions, life stories.

I recall that one of the first new waves of criticism of ecologists was that we were described as being very good at raising problems and various biological fallouts. But when it came to offering solutions, we were, at best, halfhearted. I think that since the time of Curtis, the larger environmental situation has certainly changed, as we've grown more secure about our prognostications. DDT was a bellwether once the links from Rachel Carson's critique to the actual disintegrating eggshells of eagles and eggs of other birds was made clear.

But, frankly, most newspapers still prefer, it seems, bad news over good. Bad news makes for better live coverage. An ugly divorce is easier to report upon than a happy marriage. Journalists are hard-pressed to get success stories on page one. When the NASA/JPL Mars lander, *Curiosity*, touched down on the Red Planet, no network, other than NASA's, even bothered to interrupt their normal course of programming to announce it—although the high-def imagery of Mars made top Internet billing for a day. In his book *The Adventurer: The Fate of Adventure in the Western World* (1974), the late philosopher Paul Zweig brilliantly extrapolated from the fact that even the first men to walk on the moon quickly drew a yawn from most readers, that's how fast we grow weary of new experiences and, correspondingly, feel that we have adequately atoned for our sins. It is not a very good recipe for ecological problem-solving.

But there *are* success stories, and if we don't pay tribute to them, then I think we're really finished.

PE: For example?

MT: The bigger picture is the one that has Chicken Little becoming heroic for saving the village.

PE: You really love chickens, don't you?

MT: Yes. And I'm thinking of a story I read just the other day in the *Los Angeles Times*, the return of steelhead trout and coho salmon for the first time in a century up a tributary of the Elwha River on Washington's Olympic Peninsula, following the bringing down of the Elwha and Glines Canyon dams. Here was an instance in which ecologists with the requisite passion and know-how really got it right.

I know a Korean Buddhist nun who was once speaking before a group of students, and she made the following, incisive point: You're walking down the road and come upon a man trapped in a huge pothole. You've got a very important luncheon appointment that could involve finding funds to help an orphanage. You don't dare be late for the meeting. Yet you stop everything and save that man, even at a lost opportunity cost of helping the orphanage. However unrealistic such a scenario might be, it does go to the point.

PE: Would she save a chicken?

PREDICTING BEHAVIOR

MT: I have no clue as to what the statistically average concerned environmentalist and the behavioral studies might suggest. Being a vegan may have nothing to do with spontaneous emotional reflexes. We're into uncharted territory there.

Rooster in the wilds of Brunei. © M. C. Tobias

I suspect the vast majority of the so-called 99 percent who "occupied" Wall Street are probably not vegetarians. Mayor Sam Adams of Portland, Oregon, who ordered the Occupy protestors to leave their encampment for reasons of health and safety, was very specific in terms of his being of a mixed mind. But his ultimate determination was vested in his powers as mayor to balance the safety of citizens against the message of an unbalanced economic system. He weighed in on health and human welfare in the immediate circumstances. That suggests, by way of one newsworthy example, that immediacy is more important to some people than ideology. Ideology will have more shelf life, obviously, but that's the point: immediate crises demand firemen to put out specific fires. If people stand around debating such things, the fire will consume the house, the chicken will die, or the orphanage will go bankrupt. If timing is everything, I suspect we have our answer: people act in the manner of knee jerks, not deeply thought-out, long-term convictions. Is there merit to that response and the efficiency of the response time? No question. So these are basics in terms of how we even frame the priorities, the very narratives.

PE: There are also a lot of psychological factors—and in many cases, it's known that we actually have decided to take action before we know we've "made up our minds."

MT: Isn't it all psychological?

PE: There is certainly a vast literature in ethics on such questions. Here's a classic one: You're standing by a switch on a railroad line, and there's a runaway train coming straight toward you. On one branch there are six people who will be run over if you switch the train, but if you switch it the other direction, one person who is on the tracks gets killed. And it turns out that people, in this case, usually choose to kill one to save six. But you get a different answer if the way you have to kill one is by shoving him or her in front of the train rather than throwing the switch. Gruesome but true. Of course, if that single person were Hitler or Stalin . . .

MT: There was an Iranian poll that stands out in my memory. If you have to cross the river in a boat and you can only carry so many people in the lifeboat, whom do you choose to take, whom do you leave behind? A mother, a child, a spouse?

PE: Is it the newborn baby?

MT: Curiously enough, the sample suggested it was the mother that you took on board, because you can only have one mother in life. I find this very interesting, an insight which suggests that culturally there's such a multiplicity of avenues for ethical rumination and behavioral options, that it is little wonder we're all in such a mess.

THE DEFINITION OF A HUMAN BEING

PE: Much of that mess stems from the preexisting bias of emphasis. In other words, are you speaking in terms of legal, ethical, political, or scientific explanations, codes of behavior, wisdom, practical, everyday mitigation of the chaos inherent to our existence, the "buzzing confusion" presented to a newborn that we spend our lives sorting out, and the contradictions that even after the sorting are everywhere around us?

MT: Well, it is a whirlwind when it comes to consensus over ethics. What do you find to be the real hot-button ethical questions?

PE: I've had lots of discussions over time about the idea that a human being is created and fully individual at the time of the zygote—that is, the single cell resulting from the fusion of a sperm with an egg. In my view that's like confusing a blueprint of a machine with the machine itself. Life is, of course, a continuum, and where in that continuum a new individual comes into existence is purely a social/legal decision. But I would claim that almost everybody in any culture doesn't believe that all human lives—eggs, sperm, even people after birth—have the same value. For example, if you were forced to choose whether to sacrifice a newborn baby, a ten-year-old, an adult, or a ninety-five-year-old, most people would give up the ninety-five-year-old first, and I, at least, would choose the newborn baby over a five-year-old, because the newborn is not yet socialized; the parents haven't formed the total bonds with

Family outside Mexico City. © M. C. Tobias

it they soon will and so on. But the minute you're willing to even consider that question, then you're saying that you don't believe that there is an absolute value to anything you define as human life. That's something that requires serious consideration, as does the issue of whether it is ever ethical to create situations where such decisions are necessary.

It's easy for us to agree that the Nazi eugenics and "racial purification" programs were unethical and that an abortion decision should rest with the mother or both parents. But for me, it's more difficult to decide whether it's ethical for parents to use ultrasound to determine the sex of a fetus and then abort it if it's female and you want only a son or vice versa. Does a government have the ethical right (or duty) to make such a practice illegal? What about aborting it if it will be a third child? A fifth? Does society have a moral duty to intervene with crazed over-reproducers like the "Octomom"?

MT: Let me take this to a different culture and a very different point of view. Islamic clerics, in seeking to find support for their views vis-à-vis vasectomy, which is fully authorized within the Koran, will say that certainly the soul does not emerge until after the first trimester.

PE: Well, yes, every culture harbors varying suppositions about when an individual begins, particularly amongst those cultures that also show very high infant death rates. In some cases, the newborn child isn't considered a member of the group or even real until some landmark is passed. The women, at least in some cultures in the old days, gave birth alone and were not infrequently expected to smother the infant if it was not perfect. Ethical decisions are often difficult to analyze unless you believe they're based on ancient declarations of supernatural entities.

MT: Yes. But cultures have varied with often extreme forms of hostility embedded in their fundamental belief system. Ancestor worship, in the case of medieval Easter Islanders, is believed to have manifested as so obsessive a reliance on the concordance with those deceased as to all but condemn the living. The same can be read into the record involving sacrifices of young people, particularly girls, in ancient Aztec society. In a contemporary setting, there have been documented outrages against girls in many parts of India. For example—despite it being illegal, of course—continued instances of wife burning, the so-called sati dowry deaths.

PE: Hostility toward women is almost a cultural universal.

MT: In one state in India, for example, I came across data in the early 1990s suggesting at least several thousand documented cases of newborn females being buried. Most people who consider themselves environmentalists never have to actually put themselves on the line and make ecological choices, getting by without having to be part of some categorical imperative; or espousing so-called values that are global, while failing to save the house right before them from burning down. I remember once writing about endangered species in India and Indonesia, only to be reminded that there were at least five known endangered species in my Southern California neighborhood, which I had failed to even think about.

The values we tend toward are those that perpetuate pre-existing values. It's old news in psychotherapeutic circles: reaffirmation of self-conceptions. Levels of validation tend to be egocentric and anthropocentric, and certainly in the nineteenth century, when Britain was the supposed leader of the world, its hegemony always came first, so that Britain was central on a world map. America now places itself central. Angola? I've never seen a map with Angola in the center, but I wouldn't be surprised if one existed.

MAHB: AN ORGANIZATION TO SAVE CIVILIZATION

PE: There is a proliferation of self-interest in the world when one speaks of human beings. It is predominantly a matter of religion and geography, which, in turn, spawn all the obvious characteristics of fiefdoms, regional loyalties, and priorities that have no basis in anything remotely universalist. This is one of the chief reasons I'm so committed to what we call MAHB—the Millennium Alliance for Humanity and the Biosphere.

MT: That's a whopper of an initiative.

PE: It's a new organization (mahb.stanford.edu), dedicated to mobilizing and uniting all the disparate forces of civil society trying to steer human cultural evolution in the right direction—to solve the environmental crisis. MAHB's goals include uniting efforts in academia of the social sciences, humanities, and arts with those of the natural sciences, and working to create institutions with "foresight intelligence." The latter is not just learning to routinely look down the road, but also to take action if you see the bridge has collapsed. Trying to get these together internationally with concerned citizens, industries, and governments, when there are such disparate cultures—all in a sense with different languages—is a gigantic and possibly hopeless task.

If the goal is to get agreement around the world on what to do on any of the "perfect storm" environmental issues now facing civilization, a vast array of different cultural, institutional, and linguistic orientations must be considered. The relations of the zygote, fetus, and individual are biological and evolutionary emblems of this disparate ethical array of contradictions, warring visions of how life is and how life should be.

You've got to bring together separate cultures, some of whom may believe that the individual begins when there is ensoulment at six months, another that believes it occurs at six years, and another that the magical process is instantaneous when the zygote is formed, and yet another when the fetus quickens. And, of course, some of us believe the whole idea is ancient nonsense. But the point is to preserve civilization; if that's worth doing, the whole gang must get together.

There was an interesting situation in California recently. Do you believe in involuntary genital mutilation?

MT: Circumcision?

Top: Dolls, Mexico City. © M. C. Tobias

Bottom: Edward S. Curtis, *Self-Portrait.* Courtesy of Cristopher Cardozo

Top: Melting glaciers, southeastern Alaska. © M. C. Tobias

Bottom left: Rooster at Gut Aiderbichl, Salzburg, Austria. © J. G. Morrison

Bottom right: Young orangutan, Borneo. © M. C. Tobias

Top: Sikkimese child. © M. C. Tobias

Bottom: Namibian cheetah. © M. C. Tobias

Top: Bhutanese children. © M. C. Tobias

Bottom: Wildflowers, Rocky Mountain Biological Laboratory. © M. C. Tobias

Top: High wildflower meadow, Rocky Mountain Biological Laboratory. © M. C. Tobias

Bottom: Gothic, site of Rocky Mountain Biological Laboratory. © M. C. Tobias

Top: Three young Buddhist monks at Paro Dzong, Bhutan. © J. G. Morrison

Bottom: A baby in Indonesia. © M. C. Tobias

Top left: Critically endangered Florida panther. © M. C. Tobias

Top right: Critically endangered Pacific pocket mouse with smallest radio transmitter in the world. © M. C. Tobias

Bottom: Nymphalid butterfly (*Charaxes* species) in Kuala Lumpur, Malaysia. © M. C. Tobias

Top: Farallon National Wildlife Refuge, California. © M. C. Tobias

Bottom: Heading for the hills, above the Rocky Mountain Biological Laboratory.
© M. C. Tobias

PE: Do you believe in torturing babies who have no choice in the matter, causing them extreme agony and reducing their sexual pleasure for the rest of their lives (as Maimonides almost one thousand years ago claimed was its purpose) in order to make them members of a group? It was almost put up for a vote in at least two California cities.

MT: There are going to be a lot of members of various faiths who will be very unhappy. It is not torture. I was circumcised. No problem. Perhaps—as with a gay friend of mine who was circumcised when he was eighteen and was constantly getting erections that would bleed in the immediate aftermath of the circumcision—it might constitute an inconvenience. Legislation banning it would, in my opinion, be draconian legislation and violate the Constitution. On the other hand, I totally support any bill that would, as in parts of the European Union, make it illegal to mutilate the ears of Doberman pinschers. It's difficult, in this electronic—exobyte—generation, to posit any culturally or ethically encoded universal.

In England there is a debate on at the moment about whether having mayors of cities, like that of London, are realistic approaches to good governance. Many say that no one person can possibly speak for all constituencies. If that is the case, the American or Chinese or Russian or Nigerian or Brazilian presidencies are no longer viable elements of governance. Last night over dinner, our discussion with John Harte underscored the difficulties of individualism. Anybody can aspire to become anyone, accomplish anything, and this is a divining rod of the American Wild West culture.

Yet this very power and individualism may be leading us further down a road of unsustainability within every arena, because it provides a means by which seven billion–plus people, if we were to extrapolate to current demographics globally, could imagine themselves not only capable, but privileged, to the extent that they should be granted this, that, and the other.

PE: Well, the problem of whether it's legitimate to protect babies from torture because torturing them was a habit of ancient, ignorant desert nomads is one where people obviously differ. You claim to have suffered no inconvenience when you were circumcised, but you just don't remember your feelings then. You, like me, are often inconsistent. Why then do you assume that a chick suffers any inconvenience when it is beak-clipped or a chicken when it's killed? A newborn baby is a much more developed and complex organism than a chick or even an adult chicken, as we're increasingly learning. They even recognize their mother's voices at birth. The hygienic stuff is bullshit invented to promote more business for doctors—any kid can be trained to keep himself clean. And what do you mean torturing small children is "basic" for Jews? Is that like killing Jews was basic for Nazis?

MT: Paul, wait a minute. This is beyond ethical discussion because the presuppositions are, in my opinion, without clinical or neurological substance. I know no Jew who would tell you that circumcision was painful. Perhaps you do. But I do not. Moreover, if MAHB is the key to ascertaining those points of convergence wherein there are thematic flash points, positive pathways that the majority of scientists and people in general can agree upon in an effort to improve the conditions of life on Earth—both for our species and others—then we need to focus on those areas that are not cloaked in so much fever-pitch, culturally encoded difficulty. Circumcision is one of those difficulties, and nothing we say here is going to shed valuable light on extricating ourselves out of a hole or finding useful platforms from which to extrapolate other, more important and soluble issues.

PE: All right, consider the issue of whether we all have the right to be rich and consume whatever we want. The Western focus on the individual clearly has some serious downsides—a more socially focused culture likely would be more viable in the world we're creating.

MT: If you have that expectation, then you're not going to settle for anything less than that. Which means that ethical compromise is unlikely to be reached.

PE: It means, in technical language, we're in trouble, no matter whether you are fighting ObamaCare or voted for it. Whether you are pro-chicken or eat chicken; pro–End of the World or worried about the end of the world; Darwinian or anti-Darwinian.

MT: If everyone were to maintain the high-end Western consumerist lifestyle, the carbon and biological footprint would be disastrous. We know to just supply that lifestyle to today's seven billion people would require something like four more Earths. But everyone doesn't aspire to a ten-thousand-square-foot home, and therefore you don't necessarily need the critical mass of destruction stemming from individualism. Immanuel Kant's categorical imperative is a theory, not praxis. There's always this lag time behind the ethical philosophers and ideologies and activists and actual legislation or enforcement of their goals.

But, in fact, the problem in terms of population and consumption has also taken on a new dimension that is unprecedented, historically speaking: namely, more households than ever before. So whether it is ten thousand square feet or a studio apartment, the infrastructure requirements—power, water, gas—require more and more resources commensurate with this rage for individualism and efficiency, most of which is now concentrated in cities and many megacities exceeding ten million denizens. The ethicists—those who would critique the causes and consequences of human supremacy among all other species—are outnumbered.

ALTRUISM TOWARD RELATIVES

PE: The ethicists are typically way behind the learning curve with respect to their pronouncements. There are ethicists, of course, like animal rights luminary Peter Singer, who have tried to look at the global situation. I can follow his reasoning and agree with it to the point where it tells me that I should think as much about a starving girl in Africa as I would for my own granddaughter if she were starving; but he has not yet convinced me that those are ethically exactly the same thing and that I should care equally for them. And even if I do ever get to that categorical imperative, I'm likely to continue to give preference to my granddaughter.

MT: This is another instance of making impossible choices. Once you step off that train and you look to your daughter on the left and your son on the right, and you have to make a split-second decision at gunpoint, surrounded by Nazis, in whatever form, all bets are off, I suspect, because we do not know enough about human nature to predict what we will do. And predictions, in and of themselves, do not help us. Even if we could predict—anything—would that save us?

Orangutan in Borneo, Kalimantan, Indonesia. © J. G. Morrison

Consider the lawsuits filed by the Sierra Club, NRDC, and others to contest new solar installations in Southern California in order to protect endangered wildlife, like the desert tortoise. Angry readers blog about the bigger picture. I've read some comments whose anonymous authors suggest that solar power is more important than the odd extinction. You'd think that we'd be able to quantify the impact of a certain percentage of renewable power's relationship to climate change mitigation and its sequential effects, or lack thereof, on endangered species, on an incremental basis.

We have all these scientific maps, overlays that look at every conceivable impact: low-level ozone, nitrates, sulfides, temperature variations, sea ice, whatever it is. Maps of net primary production. Yields in every agricultural sector, futures, margins, edge effects . . . Our species occupies itself with every conceivable piece of information. So why is it so damned difficult to target what we know, join forces, as your MAHB is attempting to do, and get down to doing the hard work of actually saving human beings from hunger and other species from extinction? Why aren't interdisciplinary ecologists working with energy experts and climatologists and species-specific experts, able to assess such causal relationships, and thereby improve the overall legal protections necessary to make informed decisions?

PE: Hell, that kind of coordination doesn't even happen in China, where the central government is relatively ecologically conscious. For instance, when you drive into a city like Xi'an or Chengdu, you pass dozens of high-rise apartment buildings being constructed to house the influx from the countryside, all with no sign of construction for cross-ventilation (or high ceilings), but with air-conditioning. The outside air is foul with the effluents from the coal-fired power plants that supply the energy for the air-conditioning. It is the only thing that makes the apartments livable for much of the year. And, of course, the CO_2 from the power plants is warming the planet and reducing the period in which the apartments could be used once China can no longer afford to generate the power needed for the cooling systems. On top of that, there's little infrastructure—transit systems, shopping facilities, and so on—being created around the buildings that are supposed to house some of the roughly 250 million of people moving from or being moved from the countryside in the next decade or so.

Clearly there is no sensible long-term planning among, say, city planners and ecologists, any more than there is here in the United States or most of the rest of the world. Not long ago I was lecturing in Thessaloníki, Greece, and my hosts put me up in a beautiful room at the top of a high-rise hotel. It had a glorious view of the ocean. But then the power went out in the city, and within ten minutes I had to evacuate the room—

it was on the sunny side and had no cross-ventilation, and the temperature quickly soared to dangerous levels.

THE ENDANGERED SPECIES ACT

MT: I don't disagree that local, regional, state or provincial, or federal statutes pertaining to the environment are sluggish in evolving. Similarly, international or transnational collaborations are legally far too cumbersome to incite much enthusiasm. Even a cursory look at global ecological treaties yields a fairly backward view, whether with respect to migratory species, river pollution, or nuclear weapons.

On the other hand, there have been progressive changes. We now know the Endangered Species Act has been at least partially successful. We can track gains, whether with black-footed ferrets or California condors or the above-referenced success story with steelhead trout and coho salmon. Or that in July 2012, two small Mexican gray wolf populations were discovered in the American Southwest, bringing the number of this Critically Endangered species up to fourteen known populations in Arizona and New Mexico, thanks to the efforts of the U.S. Fish and Wildlife Service.

PE: While the Endangered Species Act (ESA) has had considerable success at preserving some charismatic organisms in the United States, it has hardly slowed the crucial and more important decline of population diversity and ecosystem services. Those services include such vital things as controlling the climate, pollinating crops and protecting them from pests, generating soils vital to agriculture and forestry, controlling floods, recycling nutrients, and on and on. And the ESA has had too little impact on that.

Critically endangered Arabian leopard, Taif, Saudi Arabia. © M. C. Tobias

MT: Not to mention gun control. I'm thinking specifically of people who have taken the law into their own hands and shot wolves in Arizona and New Mexico. Those fourteen known populations I mentioned translate into a mere fifty-eight individual wolves.

PE: The Second Amendment was designed to assure that everyone could have a single-shot muzzle-loading musket in their closet, with a horn full of powder and some lead balls, in case they were needed to fulfill their duties in a "well-regulated militia." I've owned lots of guns but would be delighted to let everyone keep such a musket and some ammunition, and ban all the rest. But in practice, banning ammunition would be most effective—running down all the guns would now be a monumental task. You've got to hand it to the NRA, they're one of the most effective corporate lobbying groups of all time—slaves to "Murder Incorporated," which includes not just the gun and ammo manufacturers but the cigarette industry, the fossil-fuel-paid climate-change deniers, and many others. They kill people, now and in the future, to increase profits.[6]

And it's not clear to me that in complex adaptive systems like the interacting biosphere and sociopolitical system, more exabytes of information will solve our basic predictive problems. We know where our society is going; we just don't know how to set human cultural evolution on a course to sustainability. One problem is human heterogeneity. There's not one human nature. Everybody's different.

MT: Well, there are seven billion–plus human natures that follow upon President Nehru of India's famous line. He said we don't have a population problem in India; we have three hundred million population problems. In a couple of decades, India will be 1.5 billion. I'll never forget meeting with the State Family Planning commissioner of China, back in the mid-1990s. I was in her office, and she told me all about their fifty million mostly volunteer state family-planning commission workers in the field, including the barefoot doctors; and then she uttered the inconceivable—no pun intended—namely, that she was pretty convinced that the Chinese would hit two billion. That was a heretical thing for her to say.

PE: Now it looks like they won't. If we avoid global catastrophe, China will probably top out at about 1.4 billion and then start a very desirable slow decline.

---

6. Robert N. Proctor, *Golden Holocaust: Origins of the Cigarette Catastrophe and the Case for Abolition* (Berkeley: University of California Press, 2011).

MT: Which is a terrifying prospect given that China probably has as much or more wilderness left to lose than any other country (over 20 percent in terms of overall size). And at the rate of their runaway consumerism, they'll lose it. Moreover, the recent incident with the human rights dissident Chen Guangcheng—prior to his diplomatically negotiated arrival in Newark, New Jersey, on May 19, 2012—is that his "revelations" regarding forced sterilizations in some regions in China will only add fuel to the anti-abortion lobbies in the United States, as Planned Parenthood continues to come under fire in state after state. This is a tragedy for American women and for the planet.

PE: Although the projections for China's population, unlike some 40 percent of the world's nations, seem to be stabilizing, there are signs that the rate is going back up in some of the richer countries, which is exactly what you don't want to happen. Recent UN projections are truly frightening, especially for Africa. For example, Nigeria has about 170 million people today, but by 2050 it is projected to have 440 million people—40 million more than the projected population of the United States. Egypt, which can't feed its roughly 85 million people today, is expected to have 105 million by 2030—having added roughly the equivalent of today's Australian population in a little over fifteen years.

MT: In Mozambique, within East Africa, rural total fertility rates are at 6.6 and 4.5 in urban areas. And in the case of China, as more and more wealthy couples find the cost imposed for having more than one child to be of trivial significance overall, I imagine there will come a time, and soon, when there simply isn't enough water or food to support your estimated 1.4 billion Chinese, who are becoming largely carnivorous. I don't care how many skyscrapers of gold they build, the sky will fall. But unless the Chinese figure out a way to tax infractions of the one-child policy—a market-based, sustainability-based tax—they are all but doomed, in my opinion. I don't care how many people might applaud the fact of a mother having nineteen children and suggest that there should be no limit to the number of children one has, any more than the number of flowers. What a leap of the imagination!

Although, I should add, the Second Conference of World Cultural Forum in Hangzhou, China (May 17–19, 2013), is an impressive effort on the part of the Chinese to make meaningful strides toward tackling that country's serious ecological and demographic problems, whilst setting a fine example for what they are characterizing as a World Ecological Civilization, one based on strictly sustainable principles.

PE: Sadly, I see this as largely pie-in-the-sky—their problems in my view are monumental, stemming in part from their adoption of a growth-manic capitalist economic system. Anyway, remember the famous letter written to advice columnist Dorothy Dix: "Dear Miss Dix, I understand that every fourth child born in the world is Chinese. Oh help me, Miss Dix, I'm about to have my fourth child."

MT: Well, I know her namesake Dorothea Lynde Dix advocated on behalf of poverty-stricken, mentally challenged Americans and is credited with engendering the first wave of mental asylums in the United States. I often wonder if there shouldn't be a specialty amongst psychiatrists that focuses expressly upon family planning and upon ecological and animal rights activist burnouts, because that is certainly one of the occupational hazards.

When you juxtapose the demographic problems with the

resulting ecological ones, is it any wonder that the individual has all but ceded his/her powers of ethical decision-making to elected officials—strangers, in other words? We have become powerless to make the changes that need to be made.

And that has been the most grievous component of the sociology of power and of its abuse down through the ages. I think perhaps this carries over to our ability to be ethically assured that what we are doing stems from a personal conviction as opposed to being something we are told to do. People fight back with their convictions to a point. Some get shot doing it, perhaps protesting in southern Syria. Others get nailed for posting a position paper that gets peer-reviewed and slammed, or, worse, gets dragged through the media, and they find themselves sued for disparaging vegetables or meat, especially in states like Texas or North Carolina.

PE: The more people you have, the more government you need. So the right-wingers, who want everybody to have six kids but insist upon a smaller government, simply show with their crystal-clear insanity how little they understand how the world works. I like to use the term "population control" because it pisses people off so much, but it's obviously a major ethical function of government. In other words, if the government has the job of promoting, in some sense, the social good, it must increase the odds of a successful decline even as billions are being added to the population. We're already far behind globally on gender parity, economic equity, let alone the chances for transgenerational equity. Our world is getting poorer and poorer; one ecosystem after another becoming depauperate, depleting its natural capital, threatening the well-being of our descendants. It's the government's job to see that there are fewer descendants in the next several generations in order to give them a chance for decent lives and create the opportunity for many more descendants to exist in the long run.

MT: I mentioned the attacks upon Planned Parenthood. It's just ludicrous that we're still even engaging in such debates.

ABORTION

PE: When you ban abortion, you kill women. And the indications are that the ban doesn't prevent any abortions. The immoral Catholic bishops ought to contemplate this, especially since their "flock" has resoundingly rejected their medieval patriarchal and sexist ideas, and their ridiculous views on human sexuality.

MT: If you are actually pro-choice, then you are truly pro-life.

PE: And you're pro—your grandchildren, great-grandchildren, great-great-grandchildren. One of the things that I'm convinced of is the reason that people don't care about what happens to future generations, beyond a certain point, is that we are actually wired to associate with people whose faces we can picture. And for individuals throughout hundreds of thousands of years of hunting-gathering human history, that was a group of perhaps 150 people. Our brains make strong associations with members of a group that size, and we still have a tendency to have about that many close associates—people on our Christmas lists, for example.

In other words, we relate to those with whom we have had substantial contact. And you don't normally have substantial contact with anybody past your grandchildren. One of the

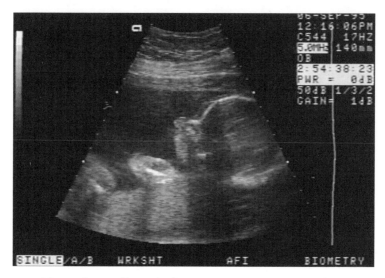

Ultrasound image of human developing in utero. © Dancing Star Foundation

strange things that's happened more recently is that we have developed huge mobs of "pseudo-kin," that is, people we see all the time and therefore feel related to, even though we're not. A million people went to Princess Diana's funeral because they saw her moving image all the time and felt like she was family.

Johnny Carson had people he didn't know approach him as if they were old friends time and again. After I was first on the *Tonight Show*, strangers used to come up to me and speak as if they knew me. They don't seem to realize commercial TV is a one-way street—that they could see me but I couldn't see them. But all of us seem to have the tendency to relate to the characters with whom we become familiar on TV shows or even in novels. If the hero is killed, you feel sorry, even if you know it's all imaginary.

Similar relating happens when you're a professor at a large university, as I have been for over fifty years. You have hundreds of undergraduate students in classes, year after year, and you may be something of a big deal to each one of them for a while, and they form a relationship with you. But it's difficult to get to know any one of the students, except in really extraordinary circumstances, say, when you have a genius sitting right in front of you, which is, admittedly, rare, or just a student in a seminar who really "gets it." As for the rest, you just don't see them enough. Grad students are an entirely different story.

BAMBI

MT: It comes down to the adage "familiarity breeds love, not contempt," as the traditional adage would have us believe.

Sociobiologist and myrmecologist E. O. Wilson says it all the time and clearly believes it. But if that were true, then no farmer would ever kill a pig or a chicken that he or she had raised. Whenever I reread Felix Salten's *Bambi*, I am both inspired and dismayed; appalled, really, that a great book can have touched so many people around the world in so many languages, but, for all we know, had very little influence on their actual diets or behavior or thinking. That enormous gap worries me, because we could talk and talk, and people could write and write and teach and teach, and legislators could enact all the laws in the world, but this fundamental stubbornness, this intransigence, this inability to be truly in step with one's convictions, in the end inhibits our ability to actually grow and mature and be more kind. That worries me, to be sure. It poses a moral dilemma that destroys my very confidence in humanity.

E. O. Wilson's latest work, *The Social Conquest of Earth* (2012)—referring primarily to termites, ants, sawflies, and non-

Deer and wild boar family in rural Portugal. © M. C. Tobias

stinging bees and wasps—suggests that we have now mathe-
matically proved that populations are pre-adapting to much
more than individual, or even family, altruism, but rather to
species altruism. I realize it has evinced some criticism. Richard Dawkins, in his article about the book in *Prospect Magazine*
(May 24, 2012), describes how Wilson's replacement of "kin selection" with "group selection" had first come out in a theory,
coauthored with two mathematicians in a *Nature* paper during 2010, which you know apparently elicited strong dissent
from numerous evolutionary biologists. Wilson responded in
a terse but effective letter basically suggesting that if new theories of science were affirmed or denied according to polls, we'd
still be following the Ptolemaic system of geocentric mapping
of the universe and the seventeenth-century postulation that
metals were oxidized by the existence of a mysterious, invisible substance known as phlogiston. Moreover, Wilson made
it clear that since that *Nature* publication, not one person had

been able to refute the mathematical theories set forth. I'm in no position to suggest that Wilson's book has gone out on a limb, but you certainly are.

But, forgetting that, if you can, I take away from Wilson's unprecedented respect for social insects and his comparison with humans a certain degree of optimism such that species altruism—as least as a concept, one perhaps as vague, yet compelling, as non-violence—might just save us and the world. Don't ask me by what evolutionary mechanisms or genes, because I have no clue. Moreover, I believe that the protection of the nest, which Wilson makes much of, is a less compelling strategy than the actual notion of kindness, particularly the feedback from kindness, which seems to me to be a very powerful force, and one that induces pleasure, not flight or aversion. What do you think?

PE: Well, it's a loaded word because again there is this salient issue of to whom are you being kind?

MT: If you have to ask that, then you'd also presumably ask, under what circumstances? An aria does not exist in isolation.

PE: If you have, as is often the case, a herd of deer that are destroying the vegetation on an island and, therefore, starving the other animals on the island, is the kind thing to do to shoot the deer and feed their meat to hungry school kids somewhere?

MT: You're positing a huge hypothetical, not to mention an unlikely scenario, although I gather there has actually been some talk of it in California. But what is less hypothetical, however, is the likelihood that those same kids, or most of them, have either read or at least heard of *Bambi*. So the contradiction is truly a human one.

PE: Yes. But there have been many actual cases of that sort—it's hardly hypothetical. It seems to me that there's got to be some kind of sensible balance, and it's very hard to get to.

MT: In our TV series *State of the Earth*, you discussed how during the Middle Ages, for example, nobody questioned the viability of initiating the construction of a cathedral, even though it might take one hundred years or more to build. Such was the belief in a stable future. Moreover, you pointed out that hunter-gatherers could simply move farther afield if they found themselves lacking for game or disagreed with their neighbors. Of course, this also invites consideration of the entire history of nomadic itineraries, as well as the forced migrations that resulted from ice ages and other dramatic changes in climate and resulting plant, shrub, and forest composition. But that "sensible balance" you mention was gone by the time in 1803 that Thomas Jefferson consecrated his Louisiana Purchase, effective a year later. Of course, for a mere $15 million, he got the deal of the century—828,000 square miles—but he also, as I understand it, knew that America would need that land as its population continued to multiply. George Perkins Marsh certainly recognized the signs of ecological overshoot in places like Haiti. And then there was the famous Paul Ehrlich/Julian Simon wager of 1980, in which Simon basically claimed resources were infinite. The famous wager, referring to price increases or decreases for five metals—copper, tungsten, nickel, tin, and chromium—was, according to many at the time, lost by you. But had the bet been made today, you would have won, by virtually every inflation-and-technology-adjusted line of reasoning.[7] With regard to the corresponding demographic changes—which were actually core to the overall debate—you were, and continue to be, proven more than right. Indeed, in my opinion, you actually underestimated how bad things would get, just as Malthus did. Your analysis in your book *The Population Bomb* (1968) was—for all the fanfare and controversy it generated—actually moderate, reasonable, even tempered, by my accounting.

WHAT IS ETHICAL?

PE: But again, it's ethics versus education. What was ethical to somebody like Simon, who in my opinion was basically ignorant of how the world works, is different from the underlying ethical arguments that trigger powerful ideas, attitudes, and hopefully solutions among those who are knowledgeable. Physicists who have worked and continue to work on nuclear weapons faced and still face a whole series of ethical choices. At one point early on, before a nuclear weapon was tested, they couldn't even demonstrate with absolute assurance to their own colleagues that the 1945 Trinity test in New Mexico wouldn't ignite the atmosphere and destroy humanity and the rest of life. They went ahead anyway, depending on the opinions of the two scientists they considered the brightest. The battle between the fascists and democracy, if you want to put it that way, was sufficiently horrendous to them that they thought it was worth taking a tiny chance of blowing up the entire world and ending all of history.

7. Paul Sabin, *The Bet: Paul Ehrlich, Julian Simon, and Our Gamble over Earth's Future* (New Haven, CT: Yale University Press, 2013).

Poached tiger, historical, Southeast Asia. © M. C. Tobias

MT: Einstein, of course, later seems to have lamented and openly questioned his discoveries, given their use for purposes of war. For his part, Leo Szilard embodied his own cautionary tale within the confines of his little collection of short stories, *The Voice of the Dolphins* (1961)—a kind of *The Day the Earth Stood Still*. And it was a forlorn endeavor because he couldn't get his message across. And it's interesting that today the Democrats, in looking at Obama's situation if you will, are probably without resolution, certainly not of the kind that led Harry Truman to make the kinds of decisions he did. I hadn't been born yet and only much later read about things like the 509th Composite Group, Site Y at Los Alamos, the whole Manhattan Project with its 393rd Bomb Squadron, B-29 Superfortresses, and the *Enola Gay*. Am I really in a position to determine what was right and what was wrong? Is anyone, among my generation, capable of really sizing up the historical context and making the right judgment call? I doubt it. And it is especially difficult after reading, say, Jonathan Schell's remarkable book, *The Fate of the Earth*, in the light of the Japanese outright rejection of the Potsdam Declaration of July 26, 1945, in which the Allied Forces outlined their non-negotiable terms for complete Japanese surrender. I believe there was no mention, as I understand it, of any said atomic bomb, let alone the likely destructive force it was soon to unleash, in that declaration. How could one have actually conveyed such force on a piece of typewritten paper? Only the actual realization—a couple hundred thousand dead and burnt and maimed individuals, the rubble, the unreal images of mushroom-shaped clouds rising with unimaginable ashes therein—could sear the reality into one's mind. Not being able to even come close to imagining such iconic horrors does not lend ethical deliberations any sure footing, especially nearly seventy years later. The Holocaust and more recent massacres from Cambodia to Armenia, and down through history, all leave us bewildered, exhausted, unclear. Or they leave me that way, in any case. Unclear as to what human nature really is, where it is headed, if anywhere, and what it all means.

Could I, would I personally, have risked my own life and death by horrible torture to try to assassinate Hitler? Many of us, no doubt, ask that question. As for Hugh Lofting, the author of the great Dr. Dolittle novels, his war poem *Victory for the Slain*, published in the United Kingdom in 1942, declared that "In war the only victors are the slain."

WORLD WAR II

PE: And we still argue over whether Truman did the right thing in using nuclear weapons. I know a lot about that extremely complex issue. I lived through it. I recently finished reading an in-depth work on Japan's plans for defending their homeland. They had more than enough aircraft in their arsenal to have inflicted many multiples of the slaughter at Okinawa. On the other hand, there's also a strong current of people who say that it wasn't the atomic bomb that persuaded Hirohito to intervene and cause the surrender against the wishes of his absolutely crazed right-wing generals. Hirohito came to that conclusion when the Russians entered the war. If the Russians invaded part of Japan, he thought the Japanese would never have gotten rid of them. And if you look at the history subsequently, certainly that seemed likely for at least forty-five years. Moreover, even after the revelations became clear about the devastation wrought by the atomic bombs, some Japanese wanted to continue the war. An attempt was made to prevent the speech that Hirohito recorded announcing surrender from being broadcast.

MT: An old friend of mine—deceased many years now, half-Canadian—was the nephew of the fellow wearing that top hat who surrendered on behalf of the emperor. My friend said that had the Soviets, rather than the Americans, come to the aid of the post-Hiroshima Japanese, Japan would have probably become a Communist-Marxist society. He once told me that when he learned to drive, he was stunned to see everyone driving thirty miles per hour. Not a mile more, not a mile less. When he asked his driving instructor about it, he was told, "Because the sign says, 'Thirty miles per hour.' " My father was just a few days short of being shipped out to the Pacific. He was an officer in the navy stationed in San Diego. My mother remembers having to put sandbags at UC Berkeley and keep the lights out in her dorm room, when they were studying for finals, and there were reports that the Japanese had launched their Fu-Go, or balloon-borne firebombs. And now we know they did: over nine thousand of them that drifted over Oregon, British Columbia, as far inland as Kansas and Iowa. This was all five, six years before I was even born.

PE: You're a child.

MT: Fortunately, I am reminded of it every day, in wonderful ways.

PE: I'm old enough to be your grandpa. The Second World War was the biggest event in Anne's and my lives. We both asked our parents if newspapers would still be published daily after the war. Since they were 99 percent war-related news, we couldn't imagine there would be enough news to keep them going. Of course, considering what passes for "news" today (think Fox "news") . . .

MT: You're—

PE: —seventy-nine.

MT: Well, as fellow primates, we're virtually the same generation.

PE: Speaking of primates. My wife, Anne, and I were in Gombe Stream National Park in Tanzania, and we heard about an incident that I have never seen reported by Jane Goodall or anyone else but us. But Anne and I were told the following: local chimp males had waylaid a female chimp from some other group, a group that was actually eventually wiped out—

primate genocide. Not the first time, obviously. In the incident, the males pounced on the female, knocked her to the ground, bounced on her, and she scrambled and got away, but she left her infant, which the males then killed, took up into a tree, and started to eat. The observers reported that these males then seemed to "think they were doing something wrong"—an inkling of guilt? of conscience? albeit too late—and one of the males took the baby chimp's body and carried it two miles and left it on Jane Goodall's doorstep. And that's a piece of behavior that I have never tried to interpret.

MT: Well, the chimp behavior affords multiple interpretations, certainly in terms of evolutionary psychology.

PE: Yes, it certainly does. That's a discipline better labeled "gene scamming." It's attributing to natural selection all kinds of behaviors that can't reasonably be explained by genetic evolution, only by cultural evolution (which, sadly, we understand less well!).

MT: I'm reminded, in a very different chimpanzee context, of that great book by Melvin Konner, *The Tangled Wing: Biological Constraints on the Human Spirit* (1982).

He had observed a chimp coming late every day to just sit and watch a waterfall and dance. Konner ascribed that behavior to a kind of mystical enchantment that animals share. But compare that with the chimpanzee behavior you reported—the notion of a gene scam, as you put it. It touches upon this chaotic universe of choices, historical avenues that have been taken or ignored, discord, inability of people to come together to make decisions that will work for the greater good, which in and of itself is a presupposition that is bashed by many for being either impractical or idealistic in some way.

PE: But it *is* idealistic, and what I have claimed repeatedly over the last fifty years is that now the only thing that's truly practical is *idealism of a sort*. If we're going to continue to be "practical" in the old sense, we'll continue to have wars. If we're going to continue to be "practical," we'll have to keep our population and consumption mindlessly growing, wrecking our life-support systems, taking from other people, and squandering Earth's resources ad lib. Today such behavior is totally impractical; it will lead to our descendants suffering incredibly as civilization collapses.

JOHN MUIR

John Muir, private collection. © M. C. Tobias

MT: What about someone like John Muir? In the late 1800s, he was a wandering naturalist, philosopher, and author, and he, of course, built a national consciousness of the beauty of the American West, especially of the Sierra Nevada. He was responsible for the establishment of Yosemite National Park, and some call him the very "father" of the U.S. national park system. Perhaps even more important was his role in the founding of the Sierra Club at the end of the century—it's still a powerful force for conservation.

PE: A great man, indeed. What else about him?

MT: He was a practical idealist. I think his meeting with Teddy Roosevelt, and the time they spent together in the Sierras, was calculated on Muir's part to a certain extent. He wasted not one moment in advocating for what he knew to be all or nothing. He understood what the stakes were. When he lost the campaign to save the Hetch Hetchy Valley from the O'Shaughnessy Dam, he was desperate. It probably contributed to his death. Now at least one editorialist has suggested that Hetch Hetchy is in better shape today than Yosemite.

But at least Muir did not lose Yosemite. When he first arrived in the valley and witnessed the U.S. cavalry having installed itself and a train track and lumber extraction and tourists starting to pour in, he knew instinctively that there had to be a system within a democracy to protect such places. That tunnel through a redwood was part of the first bona fide Occupy movement, and it happened in Yosemite as the first park superintendent proclaimed that Americans should be able to visit and stay in Yosemite in the style to which they were accustomed. Bad idea, I think. And the same year that the last of the Southern Miwok resident Indians were essentially extirpated, one million tourists visited the valley floor. About the same time, California's living symbol, the grizzly bear, went extinct, in 1923 or so. That was some year. Of course, Muir had been dead as of 1914. He did not have to live to see World War I or the extinction of grizzlies in California.

PE: He also didn't see the decline of the California condor, the San Joaquin kit fox, that big weasel called the fisher, or the little butterfly called Edith's checkerspot. I've personally watched population after population of that beautiful insect go extinct. Well, we've gotten accustomed to a paved nation. And to accommodating tens of millions of commuters and tourists, each wrapped in tons of metal and spewing greenhouse gases into the atmosphere.

Having so many people in a democracy actually worried the Founding Fathers. If you read the *Federalist Papers* and the *Anti-Federalist Papers*, one of the concerns was what was going to happen to representative government if the population continued to grow. The basic idea originally was that you and I and Anne and your wife, Jane, and your mother, Betty, and John and Mel Harte would choose someone we knew whom we wanted to have represent us in Washington. That person would go, and he or she would be a true representative for us—they came from our educational background and general situation. When the nation was born, those dispatching representatives to Washington in many cases actually knew the person sent. James Madison, as I recall, put forth in a *Federalist* paper that one person could "represent" 30,000 people. But you know now, when each representative is representing 725,000 people, it's not very representative.

MT: That's right; the erosion of democracy with the increase in population.

PE: Well, not just the increase in population. The original republican idea was that you sent the best person to Washington, where they could obtain full information and make decisions for you. That process has been eroded to the extent that we're now actually moving toward an Athenian-type direct democracy. Modern communications and fund-raising needs keep legislators always looking over their shoulders to see what the folks back home think, and money increasingly controls policy (not that this is anything new). Those trends have some really nasty consequences—especially since, unlike the situation in Athens, warmongering leaders no longer must strap on their swords and shields, as did Pericles, and go into battle.

MT: Well, if the political arena has become so diluted and there is no messenger one can trust, our views are all but lost in the mayhem of white noise—and that's the United States, with a mere 300 million–plus denizens, compared with China or India and their coming collective of, what, 3 billion? I can only hope that a grassroots collective series of changes smacks of liberal progress as opposed to the French Revolution. This summer has seen one of the least rainy monsoons on record in India. Imagine when the summer comes in the near future and there is no monsoon. Or when China hits rock-bottom, has no money for hedging empty grain bins across most of its rural provinces, while the United States starts resembling Stockton, San Bernardino, California, or Detroit, with a cascade effect of city bankruptcies.

ANDALUCÍA

MT: But if one looks at the collectives that are most successful today, they tend to be minutely populated, like traditional primate communities, other than among baboons. The first region that comes to my mind is that of Andalucía, in Spain. There you have a burgeoning organic oil and wine industry, you have tremendous ecotourism, and the numbers are interesting because, for example, in one park in Andalucía—which is actually inhabited full-time year-round by fewer than fifty people—in one year they got thirty thousand ecotourists. So the revenue generation is astounding. It works.

PE: At least as long as fossil fuels are available to transport the ecotourists.

MT: Speaking of food and drink, have our teeth evolved? I mean, is there, to your knowledge, any recognizable, morphological evolution of teeth in *Homo sapiens*? I know that E. O. Wilson has commented recently on molars and canines in our evolution, commending meat-eating as perhaps the key to our rapid cerebral advances over all other primates.

PE: I guess it would depend on where you start, but I would say the answer is, "Yes, switching more to meat likely helped our big brains to evolve, but cooking was probably equally or more important."

HERBIVOROUS DINOSAURS

MT: I've tracked the debate, at least as far back as some of the herbivorous dinosaurs, and then to *Australopithecus afarensis*, trying to determine for myself whether this meat-eating obsession is scientific or merely confirmation theory of pre-existing biases. Biases held by meat-eating scientists. I have concluded—after nearly forty-five years of thinking about it, reading up on it, observing, and living—that it is bias. Or, stated more intelligently, that the evolution of our diet is so individually dependent as to undermine any universal arguments. And even more importantly, were one to offer unambiguous research all but proving a theory of human meat consumption as somehow instrumental to our being the way we are, I would have two—I do have two—very well-tested ripostes that should clinch the matter.

PE: As you know, I enjoy meat and seafood and thought nothing about wearing a leather jacket beside our friend Ingrid Newkirk [famous animal rights activist] when we all filmed *State of the Earth* together. Of course, I didn't know you'd be seating me directly opposite her around the table.

MT: That was good fun. That's how it should be. Anyway, here's the thing. First, if meat eating was a key player in our being who we are, then I would argue it was the pronounced chord in our music that gave way to Hiroshima, the Holocaust, and the continued internecine madness that results in our horrific, environmentally catastrophic behavior, not to mention our utter indifference to suffering meted out 24/7 to all other species from which we stand to gain something: fat, grease, leather jackets and shoes, you name it. In other words, all of the worst in us. And second, allowing for a paleoanthropological consensus (which does not exist, not even close, by the way) that might add fuel to arguments in favor of meat eating as a prime mover of our evolution, I would suggest that that was then, and this is now. We're sitting here in the twenty-first century amid ecological chaos that you, more than anybody I know, have chronicled with systematic verve and sensitivity. We have the means at our disposal—both rich and poor—to satisfy our caloric requirements without the need of killing. This fact is borne out in community after community, movement after movement. I am not referring simply to hundreds of millions of animal rights activists, but to statistically anomalous tribes like the Todas of Tamil Nadu, numbering fewer than 1,500, in South India; or the Karen of Myanmar; the one million or so Bishnois of Rajasthan; the Seventh-Day Adventists in Southern California and elsewhere; or the millions of urbanized Jains from Mumbai to London to Mombasa.

I mentioned teeth because a dear friend of mine, Dr. Tarun Chhabra—a dentist and the head of a trust that helps protect the Todas from overreaching developers and lawmakers in that part of India—was allowed to work on the teeth of the entire community of the Todas. This is most likely the first all-inclusive dental record of an entire race. And do you know what he discovered? No tooth decay. And to what does he attribute this startling finding? Their more than a thousand years of vegetarianism. He told me that without a doubt, they have the best teeth in the world. I have had the privilege of spending modest time amongst the Todas on numerous visits to their hamlets. They're not vegans because they worship a significantly endangered river buffalo subspecies, one of eighteen in South India.

PE: I guarantee you it's the sugar. There are also meat-eating societies with fine teeth, and our meat-eating ancestors had few cavities.[1] Modern studies show both meat-eating and dark-green/yellow vegetable eating associated with better oral

---

1. Bianca Nogrady, "Tooth Decay Bacteria Evolved as Diet Changed," *ABC Science*, February 18, 2013, http://www.abc.net.au/science/articles/2013/02/18/3691558.htm.

health.[2] Becoming more vegetarian via farming was one step toward wrecked teeth—the key seems to be sugar consumption. And do the Todas drink river buffalo milk?

INDIGENOUS VEGETARIANISM

MT: Yes. In fact, the buffalo forms much of the basis for their spiritual traditions, agro-pastoralism, summer migrations, even temple building. Their entire priestly hierarchy, such as it is, stems from the religion of the buffalo. Yes, it is zoomorphic, probably like that of the Egyptian cult of the bull and the god Apis. Or the Sakalava of Madagascar, the Ngoni people of Malawi, or the Nuer and Nuba of East Africa, all revering bovines. Hindus worship their humped zebu and have their share of holidays devoted to the cow, which Gandhi considered key to Hinduism. Even the Brahmanic traditions owe their very stature within Hindu culture to the Brahma bull, as with the cult of Osiris in Egypt. In India there are countless ecological benefits for keeping the bovines happy and uneaten—principally their milk and their dung, the latter used for fuel, fodder, and construction material. But for most of India, the cow has sadly fallen on hard times, very hard times, indeed. Only the Toda maintain a true and undeviating veneration. They apparently used to slaughter one buffalo a year for ritual's sake, but that tradition, I am informed, has long ceased in the interests of *ahimsa*, or non-violence.

And that is my point. Even if one could argue for meat eating as somehow crucial to our being who we are, I would

2. A. Yoshihari, R. Watanabe, N. Hanada, and H. Miyazaki, "A Longitudinal Study of the Relationship between Diet Intake and Dental Caries and Periodontal Disease in Elderly Japanese Subjects," *Gerodontology* 26, no. 2 (June 2009): 130–36, http://www.ncbi.nlm.nih.gov /pubmed/19490135.

A Toda family, Nilgiris, southern India. © Robert Radin

counterargue that who we were is not really who we are today, or will be tomorrow. All those who cite the past to inform the future ignore our remarkable capacity for non-violence, for learning from the past, for change. We can do better, and voluntary vegetarianism, which we know is far better for our health and the health of the environment, is a bellwether, I believe, of where we're capable of going, in every respect.

PE: These are interesting philosophical musings, but my own feeling is that a hundred or two hundred thousand years doesn't make much difference.

MT: I'm not sure about that. Consider the rapidity with which so many invertebrates mutate so as to circumvent humanity's most devious and ingenious pesticides. Just last week data was published regarding the fact that cockroaches, whose generational life spans are three human years, have—by evolutionary adeptness, it would appear—given up their sweet tooth in order to outsmart the poisons, mixed with sweets, to rid human kitchens of cockroaches. It was the small German cockroach that was studied by a group of Italians.

THE FOSSIL RECORD

PE: One of the things that's crystal-clear to me from the fossil record is that cultural evolution started out really slowly and then speeded up. If you look, for instance, at the primeval tool kit of our forebears, it stayed the same for something like a million years. And then the next one came along and lasted for approximately six hundred thousand years. And here we are, we're only, what, twenty-five years into the Internet culture and serious discussion of vegetarianism, and things have just not come to any new equilibrium—and the acceleration of change is extraordinary.

MT: But is it really? Has the rate of our synapses being fired off escalated in proportion to the Internet and the deluge of information? Have our brains increased in cubic centimeters? I have not read anything to that effect. In fact, a recent scientific study revealed that ants, for example, use something like 16 percent of their brains for important skills, whereas humans use a third of that, at most. And what I have studied quite thoroughly is the research concerning the impact of fire on our cultural and morphological evolution. Experts at Johns Hopkins School of Medicine many years ago pointed to the oldest-known fire pit ever discovered, Chesowanja, and nearby at a site (also in Kenya) known as Koobi Fora, which I discussed in my book *After Eden: History, Ecology, and Conscience* (1985). About 1.3 or 1.4 million years ago, East Africans sat around a campfire. Ever since then, *Homo erectus* throughout Africa and Asia enjoyed, or perhaps "endured" is a better verb, a rapid expansion of the size of their braincase. It was the fire that enabled our ancestors, who previously had held to a rather strict circadian rhythm of going to sleep when it got dark, as most, not all, primates do, to sit around exchanging information—campfire gossip. That may have done more to advance our evolution than any other single component. I can just hear them now: "Hey, check out that sunset!" Naturally, the dating of a campfire, versus a lightning-induced bushfire, has created its own storms of scientific controversy, especially in light of the proposition that *Homo erectus* was not exactly an innovator; that their tools were basically good for about fifteen minutes of cultural nuance; that other signs of genetic adaptation were what prompted the so-called big bang of mental evolution, and so forth. One of the best papers on the subject is from 2006 by John McCrone.[3]

PE: You could also argue that fire provided our ancestors the means of cooking meat as well as grains and other vegetable food, enriching our protein intake and thus accelerating our metabolic range and acuities. It likely made it easier to fuel energy-hungry brains and gain our stranglehold on the world.

MT: But many would argue that the need for protein from other animals is a pernicious myth. That we actually get our proteins from any number of food sources that are non-animal. And if you look at the myriad of early hominid teeth, it's clear

3. John McCrone, "The Discovery of Fire," *Dichotomistic*, 2006, http://www.dichotomistic .com/mind_readings_fire.html.

that the transition from herbivorous roughage to animal flesh yields no clear linear or geographically certain itinerary.

PE: You can get protein, there's no question, but you have to be careful. Otherwise you won't get the right mix of amino acids. You also need to be very careful about vitamins B12, A, and D. And getting an adequate diet from only vegetable sources involves one hell of a lot of chewing! The clearly herbivorous australopithecines were the ones that *didn't* leave a posterity.

MT: Yes, you have to think about your diet. That is, if you care about your own health. I, personally, am not concerned by it. And would not live my life strictly for myself, if my health were truly dependent upon hurting others. I'd let go, of myself, that is.

PE: I find it not credible that our ancestors, who we know scavenged and hunted, would have made exceptions. And I have no ethical problem with eating meat, even though I do so by inflicting some "unnecessary" suffering. Even sheep have been known to eat baby birds, did you know that?

MORAL EXCEPTIONS?

PE: Is the idea that our ancestors would have made some kind of big moral exceptions when they were hungry plausible?

MT: Well, we do know that there have been vegetarians over the past few thousand years. Certainly amongst pre-Socratic philosophers like Pythagoras. But whether it's valid to conjoin their example with any evolutionary hypothesis is something else. Especially if you are speaking in a time frame encompassing hundreds of thousands of years. But that's not my point. I'm speaking about now, today, tomorrow, the day after tomorrow. We are a new species if we but choose to be, and all our evolutionary advances being predicated, allegedly, upon our need for animal-derived protein, in my opinion, merely constitutes old news and has ceased to be relevant. Because I believe we are capable, as I mentioned, of change. Evolution does not condemn us, not anymore. Only our choices can do that.

And even if one does subscribe to E. O. Wilson's new position on inclusive fitness encompassing our entire species, all the more reason to posit the prospect of non-violence all-inclusiveness. I realize it's a dream; it is philosophically even slightly quixotic. But so what? Martin Luther King Jr. was quixotic. So was Gandhi. Mahavira. Susan B. Anthony. Thoreau. Even Teddy Roosevelt, the big-game hunter, got the wisdom of the Grand Canyon, Yosemite, the U.S. Fish and Wildlife Service with its protection, initially, of Pelican Island in Florida. Of protecting endangered species. We've changed significantly.

PE: I'm not saying you can't be quite successful as a vegetarian. But if they get hungry enough, I suspect most vegetarians will eat meat, and if history is a guide, even human meat.

MT: But you're taking the exception, not the rule, or certainly not the Golden Rule. You're inviting what I would call a categorical impairment or iteration of bad habits; not the formation of a new paradigm, which I believe is essential, and also within the cultural arsenal of our genetic, or cognitive, latitude. We are capable, now, of anything: Holocausts, or *The Marriage of Figaro*.

PE: Yes. But even within the survival strategies during terrible, unspeakable horrors, like the Bataan Death March perpetrated by the Imperial Japanese army, those thousands of dead American and Filipino prisoners (a few survivors are still alive) resorted to every conceivable means of escaping death. During their eighty-mile trek through hell, some tried eating grass raw, and they tried making soups out of it. Of course they ate every animal they could lay their hands on. And the smart ones actually ate the weevily rice, because the weevils were higher in protein content.

MT: I can't dispute that or would not even care to. But again, you are citing an exception, not the rule. Human history is not woven merely of death marches and killings. Discord haunts us, absolutely. Historians of warfare—from Herodotus and Thucydides, to Edward Gibbon, Paul Kennedy, and John Keegan—have not spared us the least detail. William Shirer's *The Rise and Fall of the Third Reich* (1960) is pretty clear. Still, equally compelling histories of harmony reignite our prospects to do right by one another and all other species. I'm thinking of St. Francis, of Vermeer, of the *Moonlight Sonata*, or of political reigns like that of Shogun Yoshimasa, who ended up renouncing his position in order to contemplate the roses and his sand gardens in Kyoto.

THE PAST IS OVER

PE: It's true. The past is over. We're here now, and we'd better damn well make our ethical decisions. It's just that whether it's ethical to eat meat is pretty far down my list when several billion human beings are not well nourished, billions have no real freedom, women are almost universally given the short end of the stick, many people believe skin color is an index of quality, and human behavior is destroying our life-support systems. There are good health reasons today for cutting down meat consumption, and good ecological reasons if people are determined to see how far Earth's carrying capacity for *Homo sapiens* can be stretched. But the ethical issue is tough. There are, remember, people who think it's unethical not to take great care to avoid stepping on ants.

MT: So assuming we are challenged when it comes to learning the lessons of our own history, as philosophers like Heidegger and historians like Toynbee have challenged us to do, at least we have the hope of learning our lessons today in terms of what's directly in front of us.

WHY ARE YOU A VEGETARIAN?

PE: So why are *you* a vegetarian? What lessons from history are pertinent here?

MT: Because I am expressly of the belief that meat eating by humans is totally immoral. There is no evidence, none, that people will die if they do not kill other animals. That we cannot survive without killing other animals. You yourself have mentioned the advance in artificially produced meats. NASA has long been an advocate for research into this field. But that is merely a transitional "get over yourself" kind of ruse, if you will. The reason so many vegan restaurants still use real animal names on their menus: vegan duck, vegan chicken; tofurkey (meaning tofu prepared to taste like turkey, or the real thing, so to speak). But the "real thing" is not the thing we should be eating. The real thing has feelings just like us;

Bovines at a sanctuary for rescued animals. © M. C. Tobias

mammals all nurture their young. So who are we to invade their territory, torture, and slaughter them? Mothers just like our own mother, if you want to be expressly logical and carry the argument through to its outcome.

Hence, the emergence over thousands of years of ovo-lacto vegetarians (those who do eat eggs and/or dairy products, but no fish or meat), vegans—who will consume nothing that "has a face"—vegetarians who will not eat meat but think nothing, perhaps, of driving in automobiles with leather seats . . . these are all, in a sense, comparable with the great book *Varieties of Religious Experience: A Study in Human Nature* (1902), by William James, the Harvard provocateur often thought of as the father of American psychology. It's no small coincidence that the source of the book was a series of Gifford Lectures on natural theology that he gave in 1901 and 1902 at the University of Edinburgh, since it was that very university's medical school, in the late 1700s, that first all but outlawed dissections and made vegetarianism a veritable science within the school's curriculum on the basis of morality, as documented by the so-called Oxford Group and ethicists like Richard D. Ryder.

I'm actually vegan, but to answer your question, why, truly why, I guess it's because immorality for me has a host of tactile nuances that are, at least in my case, truly hard-wired. When I walk, when I look at everything in front of me, I actually do look because I try to avoid stepping on an ant, literally. If I see a bird, which I do everywhere fortunately, the last thing on earth I would think to do for any reason would be to harm it. It's hard-wired in me, I'm certain of it. Now that doesn't mean I know what I'm doing, necessarily.

PE: What do you mean?

THE TARANTULA HAWK—FAMILY POMPILIDAE

MT: Once in Malibu I was up on a ridge that I like to climb, and I was with some visitors from another country—friends whom I was showing this beautiful place to—and I saw something that totally baffled and terrified me. I didn't know what was going on. Right there in front of me, in front of all of us, was a wasp (often referred to as a tarantula hawk wasp) and a tarantula having a fight. Well, I'd never seen this "event" in nature, if you will, and I was horrified, and the first thing I thought to do was to race down and grab a stick and separate them. That was what I did. I was silently outraged by what I was seeing, because I thought, *Oh my God, that poor wasp is being killed by that giant tarantula.* So I separated them—only, to my horror, to see the wasp race back on the ground to the tarantula, flip it over, sting it, and then drag it paralyzed away. I photographed the whole sorrowful episode.

PE: I don't find it "sorrowful." A lot of baby wasps were going to be fed. And as far as I can see, your argument for being a vegan boils down to "I think it's right," which in my view is perfectly legitimate, since what people think is "right" is the basis of *all* ethics. But I hardly find it persuasive. Consider all those who disagree with us that a woman should be able to decide whether to have an abortion and are utterly convinced that abortion is "wrong." Others may claim that their pleasure in hunting overwhelms the pain a deer suffers when it is shot, especially when that pain may spare it the torture of being devoured alive by a predator. I find it impossible to apply a moral calculus to the tarantula and the wasp, and equally tough for me and a trout I devour.

LIONS AND WARTHOGS

MT: I suppose there is an underlying supposition that exits the human moral context and enters the world as a biological principal, namely, feeding the hungry world, in this case, the hungry tarantula wasp world.

Tarantulas, like hummingbirds and army ants evidently consume an enormous amount, and range, of prey species and/or nectar. They carry huge metabolic loads. So, in my defense, I claimed ignorance (at the time). Nonetheless, my friends leapt forward saying, "Oh, this is terrible. It can't be allowed." And I didn't say anything, and I'm still thinking about those kinds of things. So it's not that I would argue that it was wrong for me to try to separate them—that was my instinct . . . learned, nurtured, genetically programmed, psychologically morphed, a new nature declaring itself in my guise, whatever you want to call it. Some might say I was just being stupid, illiterate, utterly unconscious of the primordial

Rescued lion at Marieta van der Merwe's Harnas Wildlife Foundation. © M. C. Tobias

requirements of nature. But I do not subscribe to Hobbes's version of nature "red in tooth and claw," or however he said it. Even though I have read that grotesque scene George Schaller describes when he observed for long minutes one night as a lion dragged a large warthog out of its hole and devoured it piecemeal with the pride. Sheer agony on the part of that benighted mammal, the warthog, *Phacochoerus africanus*. A grazer, a pig family member, an omnivore, known on occasion, when there is drought, to feed on carrion but never to kill.

Oh, and by the way, that trout of yours? I hope it was not one of countless rainbow trout placed in lakes and streams throughout the High Sierra to please the fishermen. These are the same bio-invasives wreaking havoc on native species, especially the frogs, who are also vulnerable to the *Batrachochytrium dendrobatidis*–induced chytrid fungus. Anyway, I was speaking of lions and warthogs.

PE: Warthogs eat worms and birds' eggs, and I suspect nestlings if they come across them. Believe me, in the right circumstances a warthog would kill and eat you sans compunction.

MT: I have been with warthogs all over the world, from Poland to Hawaii, in the wild. I love them. Schaller's description has forever given me nightmares. And, frankly, the little research I have done into the numbers game across the Serengetis of the world suggests that for every mammalian carnivore, there are probably on the order of four thousand herbivores. This ratio was the result of an initial overview, on my part, of the available literature regarding the number of estimated mammalian carnivores versus non-carnivorous vertebrates (excluding avians) in Amboseli National Park. Hence, the data is not data, but a first, albeit highly suggestive, *guesstimate* on what is actually going on out there in nature's evolutionary killing fields, so-called.

PE: I suspect that is a high figure. It probably can be checked.

MT: No doubt Schaller would have other data. It may be that some of those herbivores are actually omnivores, like grizzlies, who spend most of their year grazing on vegetable and fruiting matter, only gorging on salmon when the anadromous fish are in season, like July, in Alaska. And I will add that my tally of one in four thousand was a non-scientific calculation, an observation of counting and calculating over a period of about ten days beneath Kilimanjaro. And it was based upon vertebrates, not invertebrates, obviously. Will East Africa prove to be different in this respect from, say, Borneo? Or New Guinea? Or any of the neotropics? Or boreal forests? I have no idea. It's known that large obligate carnivores, like Serengeti lions, need something like fifteen pounds of meat per day to be healthy. That meso-carnivores—such as a fox or coyote, get roughly half to 70 percent of their meals from eating animals, whereas hypo-carnivores obtain more than 70 percent of their meals from non-animal food. You can study the evolution of molars and incisors all you want, but it won't tell you anything about the composition of food consumption in a landscape-size geographical area. Go back 400 million years, and it gets especially complicated. Among mollusks, dating back to the Cambrian period of the Paleozoic era, 500 million–plus years ago, and today comprising nearly one-quarter of all marine species, the majority are quasi-herbivorous. They eat algae, neither true plant nor animal, making this an especially complex debate, I'd argue.

PE: And if you were a tarantula wasp who adopted an herbivorous lifestyle, your evolutionary line would die out. And by the way, I've watched Toklat grizzlies dig out and devour harmless little herbivorous mice—like many other mammals they'll take meat if they can get it (not that that proves anything at all—the old "is-ought" problem).

WE'RE ALL HYPOCRITES

MT: Well, so there you go. And the world might be a better place for it, or not. In any case, I haven't been confronted with that dilemma—although a John Harte might rush to add, "But, Michael, don't you realize you drove to Crested Butte, you used up X number of gallons of gas/oil, which translates ultimately into X number of dead organisms on the planet, because you've contributed to that thin layer of petroleum on the surface of the oceans; to the fouling of everything, so you're a hypocrite and you don't even realize it."

PE: Or you will die, or others will die, from the climate disruption you have contributed to. And those others may constitute a wide variety of organisms: not just those peaceful herbivores, like the zebras and butterflies you are so fond of, but nasty carnivores like cobras and ticks that you dislike—but also lots of plants that you devour without a thought for their intricacy, beauty, or monumental billions of years of evolutionary history.

MT: But I do realize it; we both realize it. It's just that realizing it for me doesn't detract from my ethical resolve, I suppose you could say, not to eat animals if I can avoid it. That's why I'm speaking of it today, among humans, not non-humans like tarantula wasps or lions or grizzlies, when I bring up the subject of diet and morality.

PE: Well, let's put it this way. I generalize the whole thing to say that you are "eating" animals, likely including some human beings, when you have a child, feast on asparagus, or drive a car. All those activities, indeed virtually *all* of our activities today add greenhouse gases to an atmosphere already too full of them. And that endangers all future life-forms, including our descendants.

RIGHT FROM WRONG?

MT: Actually, if you extrapolate the ecological footprint fully, it transforms quite readily into a compassion footprint. The food we eat is depriving somebody else of even the most basic staples. We know that. But to get back to the ethics question: that anecdote you described from Tanzania, the chimp bringing the dead infant chimp to the doorstep of a great behaviorist, who had been there to study, decade after decade, chimp behavior. Was that a moment of conscience, or ethics, or just a strange looming sense of guilt? What was it?

PE: Well, it could not be ethical, because my definition of ethics is shared values, and chimps lack the language with syntax required to permit the sharing of values. But whether speaking of chimps or of humans, there's no external source that tells us what's right or wrong. As a social animal, we're just like you and me right now, sitting here discussing what we personally think constitutes right and wrong and what our reasons are for these viewpoints. Obviously, over a gigantic range of issues, we agree. We may actually agree on every issue, but again, on the issue of whether it's right or wrong to eat animals, I look at that—at least between you and me—as a personal choice of what makes us feel right or wrong in a certain situation.

MT: Interesting. I disagree with your definition of ethics as "shared values." I believe a solitaire can hold to personal convictions, and I call that ethics.

VEAL

PE: Well, if you ask me whether it's right or wrong to put a pig or a calf in a tiny little cage or crate to make sure its meat is tender—the story of veal and so on, well, I tend not to eat veal because, even though my own personal opinion is that a calf doesn't suffer as much as you or I would suffer, seeing the calf in that circumstance affects me negatively. In other words, the things that would make you or me extraordinarily unhappy in that situation don't necessarily apply to the calf. But on the chance that they do, I'd rather not eat veal. It just seems to be wrong. But we make our own judgments on the basis of how we feel in a situation. In other words, you might feel sorry for animals penned up in a zoo, but if you're penned up in a zoo and fed every day, you don't have to worry about the lion coming and eating you, as happened to that warthog. It may be that the animal in a zoo is in a lot better shape, and much happier, than he or she would be out "free" on the Serengeti.

MT: It's true that animals have a hard time, as far as we know, committing suicide, although there are reported incidents of it. My point being, they usually don't have the luxury of being put out of their misery with an injection. And we know from countless examples that at least some species, in some cases, may live longer in captivity if well treated than they will in the wild. Others, definitely not. And the quality of life in captivity is everything. A wild bovine may live twenty-five years. In captivity, only a few years, depending, again on what sort of captivity. I'm not sure about invertebrates. I have never kept a pet ant isolated from its group. I can't imagine experimenting in any manner on living beings.

A THEORY OF MIND

PE: It's not clear that many animals would view quality of life in captivity as "everything." You or I could get bored even if we were safe and well-nourished, but safety and satiation might be the only definition of quality for a cow.

One of the things that make people close to unique is that we have a theory of mind—we know that other human individuals have knowledge, thoughts, intentions, and so on. There are other primates, and maybe some other animals, that do so as well. Thus it seems reasonable to think of another per-

Rescued bovine at Gut Aiderbichl Sanctuary, Salzburg, Austria. © Gut Aiderbichl

son, or a bonobo, as bored. And if I thought cows or sparrows or other animals have comparable minds to people or bonobos, then I might think of them being bored. But mostly I don't. And most animals, as far as we can tell, don't have a theory of mind and lack empathy. They can't put themselves in other individuals' paws.

MT: How do you know that?

PE: There's no sign that the lion dragging the warthog out of its burrow thinks about what's going through the warthog's mind as it's being eaten. And of course we don't really have a clue what's going through its mind—the warthog's, that is—either, although we see signs we interpret as the warthog's great distress (which might just be a behavior that could distract the predator or attract its competitors, adding a slight probability of the victim escaping).

MT: And it was precisely because the Nazis and the Japanese who forced those more than 75,000 soldiers on a death march knew precisely what was going through the minds of their victims, that humans have a unique problem. One could argue—and must argue—that Descartes' mechanistic theories and his own actual experiments on animals were the workings of a sick, deluded individual. Animals are not machines. Descartes was supposedly brilliant. In my opinion, he was a maverick with no heart. A sick individual, a product not of his time—because his time witnessed such great individuals as Monteverdi, Shakespeare, and Cervantes—but of his own devilish lack of feeling.

Look, in your own book *The Machinery of Nature* (1986), you wrote, and I quote from page 254, at least in the edition I have, "Ecologists usually consider, as a rule of thumb, that about 10% of the energy that flows into one trophic level is available to the next. Thus, if green plants in an area manage to capture 10,000 units of energy from the sun, only about 1,000 units will be available to support herbivores, and only about 100 to support carnivores. This means, for instance, that roughly 10 times as many people can survive eating a corn crop as can survive eating cattle that have been fed on the corn." So there you have it. Evolutionary lines, like those of the tarantula wasp, pose greater ethical dilemmas for me than do people who eat corn soup, and who—theoretically—should more readily pass along their genes than tarantula wasps.

PE: Descartes and his helpers were certainly cruel, and I wouldn't doubt their behavior would be considered psychotic today. But so would that of the many prelates who burned people alive because they didn't follow rules handed down by an imaginary supernatural entity. And tomorrow humanity may conclude that those who genitally mutilate helpless babies because of rules passed down from ignorant ancient nomads are also psychotic. And I see no reason to declare that tarantula wasps should die out—if they did, we might be up to our butts in tarantulas, and few except (possibly) for tarantulas would love that. In addition, a superabundance of tarantulas could upset some ecosystems, since they can be important predators on many large insects and, in some places, small birds.

But as you consider any scientific explanation, you always have to consider the chance that it may be wrong. The assumption is that we're averse to pain because we think about the pain, but it may be that the aversion doesn't translate into horror in an animal's brain, it just causes the creature to go in a different direction, to flee the predator. The creature may just be programmed to escape or struggle under certain circumstances and doesn't carry with it any sense of agony, any more than a moth—a male moth—following a pheromone trail through the air to a female is horny in the same sense you would be following a drop-dead gorgeous starlet into the bedroom.

MT: Though some butterflies will perch and wait for the female—they're definitely aware of timing and opportunity and strategy, as you know probably better than any lepidopterist in the world.

PE: But, no, your word there, "aware," I wouldn't use that word.

MT: So use a different word. It's just you and me speaking. I doubt any butterflies are listening in at the moment. And the ones that may be are probably either dorsally or laterally basking to stay warm; regulating their body temperatures. You can say this is inherited behavior or learned behavior. But I'm speaking about behavior when I use the qualifier "intelligence." It's the same with spiders. New data suggest that tarantulas display certain mammalian-like behavior; that jumping spiders are particularly intelligent, and—most compelling of all—that even the tiniest of orb spiders, for example, can still accomplish feats of web construction that are wondrous and equivalent in complexity with spiders with much bigger brains. It all leads to question marks and dangling modifiers in terms of what really constitutes intelligence, feeling, imagination—all those allegedly higher aspects of behavior we credit ourselves as monopolizing.

Not to mention the fact that the tensile strength of spiderwebs is extraordinary and also acoustically uncanny: spiderweb thread has been used for the strings of a violin to great musical affect.

PE: Butterflies respond to a set of stimuli in a manner preprogrammed for a certain situation. In other words, there's no "thought" or "emotion" in our sense—or at least no indication of it. A chimp can be aware of where some food was put by reading clues, and some other animals can as well. You can sort of hypothesize awareness. But aware in the sense that you or I are aware of pain, for example, I doubt it very much. Most of the problem with pain, from my point of view, is remembering or anticipating it. The pain itself, unless you're being tortured, is—I don't know how to put it, it's very difficult to describe, and that's one of the problems we have, but it tends to be tolerable. I normally skip anesthesia when I have a colonoscopy—the interest in traveling through my own gut with a camera (but no gun) overwhelms the little crampy pain.

But pain certainly increases our fitness in some situations, such as when we jerk our hand back after touching a hot iron.

MT: Well the concentrated feedlots where you have these, you know—

PE: Animals forced to stand day in, day out in their own wastes.

MT: Billions of them.

BUTTERFLIES VS. BIRDS

PE: Those concentrated animal-feeding operations seem to me a bad thing in many dimensions. I'll give you a personal example. I don't like killing butterflies, but I've killed a lot of them in the course of my life and my research. One of the reasons I work with butterflies is that the idea of doing the equivalent killing of birds would be much worse for me.

I think birds are more complex than butterflies; their behavior is more complex and more subject to alteration, and birds tend to live much longer (so in a sense they have more to lose). It depends a lot on the bird. I mean—an English sparrow is one thing, a parrot or crow is another. Parrots and crows are a lot smarter than English sparrows.

MT: Pigeons, New Caledonian crows, tool users. Ask any really open-minded and well-traveled ornithologist, and you'll be blanketed in a deluge of superlatives that cuts pretty generally across all avifauna, regardless of size, geographical location, or what have you; of relative indices for determining, for relating to what a human would call complicated. So it opens up this realm of ethical relativism. In your case, butterflies

A white peacock butterfly (*Anartia jatrophae*), a nymphalid distributed from the southeastern United States through much of tropical America. © J. G. Morrison.

versus birds; and, more difficult still, parrots versus English sparrows.

PE: First of all, I don't like wrecking anything that's organized, that has evolved, and so I don't get any particular pleasure out of that, any more than I would by shattering a natural crystal.

MT: And if there were an alternative to killing the butterfly?

PE: Let's put it this way, for every butterfly I have killed, I have written a number on and released probably five hundred butterflies. The only time our research group has killed butterflies is when we have been doing genetic work where you couldn't do it on a single leg or something like that. Nowadays, you take a single leg off to examine an individual's genome, which doesn't inconvenience them significantly. We know that because we've marked and released many thousands of butterflies in various conditions, and when you take a leg off, you recapture the butterfly just as frequently as when they have all six legs.

MT: If I put my mind into the mind of a butterfly, well, I couldn't kill one. I applauded the fact that a group of ornithologists in northeastern India (Arunachal Pradesh) recently discovered a new species of babbler without the least invasiveness.

PE: They didn't shoot it?

MT: No, they sampled its genetics from a dropped feather and were able to determine that it should be considered a new species.

PE: Techniques change. Around 1898 the California Academy of Sciences sent an expedition to the Galápagos, and on one of the islands they discovered thirteen tortoises of a distinct subspecies that people thought had gone extinct. So they collected them. Even for that day, I would call that unethical stupidity.

MT: The New Zealand syndrome, the Walter Bullers of the world who, when they heard that a certain bird species was down to its last few individuals, would dispatch whomever they could to go capture a "specimen," so-called (or "speciwomen"), bring it back to be kept in captivity, or, more usually, preserved, painted, stuffed in a museum cabinet. There's this guilt that accompanied much of the ecological sciences, as you know, in the late nineteenth century and reforming techniques. The huia (*Heteralocha acutirostris*) was last seen December 28, 1907, in the wild. They went extinct; a sexually dimorphic large wattlebird, with incredibly distinct beaks—the female's being long and curved, the male's short. The painting by Johannes Gerardus Keulemans and those [specimens] that remain stuffed in museums are all that is left. But it was the bounty hunters and their patrons who really accelerated the extinctions.

RATES OF EXTINCTION

PE: Although, on the other hand, they would argue that species have always gone extinct; it's part of the natural process that produces new species that are useful, and therefore they're actually helping nature along. Of course, that's bullshit, since the problem now is the extremely high rate of extinction—hundreds of times what prevailed through most of geological history.

MT: Well, we get hit with that all the time. I'm sure when you wrote your book *Extinction* with Anne, there must have been countless individuals who asked you, "Dr. Ehrlich, would you not agree that evolution is continuing, and for all the many extinctions, there are new species being found every day?"

PE: And the answer is—it isn't a problem of extinction, per se, but, to repeat, it's a problem of rates of population and species extinction and a problem of high rates when we have *Homo sapiens* now, which we didn't have sixty-five million years ago, at the time of the last big high-rate natural extinction episode. Now we have more than seven billion big primates trying to live and getting all kinds of value from its other living companions, including aesthetic value and ethical value, and so

The skeleton of an extinct elephant bird from Madagascar. © M. C. Tobias

on. And as I said, the big problem, too frequently ignored, is *population* extinction.

MT: But you're understating it. Based on everything I've experienced in my life and thought about and observed, if we are, indeed, witnessing the sixth spasm of extinctions in the annals of human biology—in the annals of biology, period, I should say—then is it not incumbent upon us to leave the lightest footprint possible and to advocate for a light impact, rather than the inordinately destructive one that most characterizes our notorious collective, in view of what's happening all around us?

A DIFFERENCE IN DEGREE OR KIND?

PE: I actually feel very strongly about the inequities represented by at least three billion people on this planet not getting to live the kind of lifestyle I get to live. And I have a certain amount of guilt about that, but I don't have enough guilt to reduce myself to the average or below-average level on the planet to do what some Buddhist monks might do. I'd much prefer a human population of perhaps 1.5 billion, with everyone able to live a lifestyle more like yours and mine. And such a situation should, in the long term, allow more *Homo sapiens* to live and enjoy life, since we could probably avoid the great collapse of civilization toward which we're now headed. And it would also allow the rest of nature to persist and continue evolving—even the jungle fowl that could carry on the chicken lineage if people stopped propagating them for meat and eggs.

A DIGAMBARA ECONOMIC MODEL

MT: I'm constantly thinking the same thing about myself and wondering, particularly with a view toward what I call the Digambara economic model, Digambara being the naked Jain monks—there are very few left. They are the ones who have absolutely no possessions.

PE: They live by begging?

MT: They don't beg. In fact, they take nothing if it in anyway imposes on others. What they literally are doing is putting their hands out to householders for whom it is a rare honor to be able to provide pure food for the monk who eats maybe once every other day. It's the ultimate example, if you will, of a human being who is absolutely possession-less, has no footprint to speak of. They are the true environmentalists. Some are also activists, though Jain tradition has shied away, by and large, from anything smacking of intervention. About your point regarding extinctions, Jains are wont to say, "Live and let live." But it has a myriad of meanings.

PE: But could these Jain monks and nuns survive if the other people didn't do the things they won't do themselves? I'm unimpressed. I think true environmentalists recognize that all organisms have "negative" impacts on their environments. True environmentalists want to design a world in which they can lead pleasant lives, but where their numbers and footprints are small enough to keep society (and necessarily "nature," which supports society) sustainable. Living a naked life with no possessions is just trying not to be human.

MT: Nothing wrong with that. As far as I'm concerned, they are the ultimate environmentalists. Some Jain monks have starved themselves to death. It's called Santhara or Sallekhana or Samadhi-marana. It is *the ultimate human act*; saying "no" to a brutal world.

It may not seem relevant to any practical ecological discussion, but it is certainly an ethical consideration, relatively speaking. What if we were swept away by an avalanche and left for dead, but were not dead? Would we resort to cannibalism or simply allow ourselves to starve to death, respecting the other? In fact, at one level, as you previously indicated, half the world is, in so many words, starving to death, or certainly living near, if not below, the poverty line. So are we not all confronted by that choice, that ethical determination, as we face potentially ten billion Western-style aspiring consumers on the planet, when everything we know about the Earth's current state tells us clearly that the planet's biological regenerative capacities are being gutted?

PE: Obviously, the super moral thing to do by those standards would be just kill ourselves, and then we would take our pressure off. But unless another five or six billion joined us, it wouldn't do any good at all (the "drop in the bucket" response to such ethical imperatives), and you would lose our voices for changing the behavior. And I admit to being selfish. I only have one life to live, and I'd like to make it as pleasant and meaningful as possible before I'm decomposed. It gets a little complex, but I find the Jain route neither moral nor (for me) very compelling.

3 THE CHOICES WE MAKE

MOVING FORWARD IN SPITE OF CONTRADICTIONS

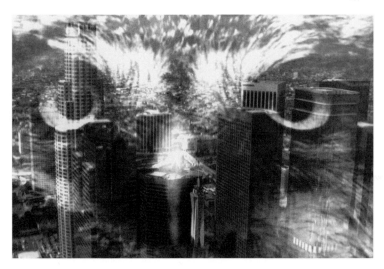

Owl superimposed over downtown Los Angeles. © Dancing Star Foundation

PE: But that's an example of personal choice—eating food containing insect parts (unavoidable) doesn't worry me except to the extent they represent poor sanitary conditions, and I would certainly not practice brahmacharya (a Jain principle of skipping sex, the most fun thing evolution ever invented).

But those choices are certainly ones I and most of our society would not consider immoral, if perhaps silly. Then maybe I'm just a greedy son of a bitch . . .

MT: Oh sure. You professors are right up there with the big hedge fund managers. But I do agree that we are all greedy to the extent that we want food when we want it. We need sleep when we need it, and we'll take it, if we can, on the best available mattresses. We are both wearing shoes rather than going barefoot. We're well equipped; I've got all this technology, et cetera, et cetera. So, I mean, do we have to set an example of the ultimate in order to be persuasive? Al Gore was criticized for flying around the world so many times and showing up in a limo, I am told (I have no proof), at one of the big premieres of his film, *An Inconvenient Truth*. I don't think the former vice president's itinerary or mode of travel nullified his urgent message. Sorry. Nor that of the 50,000 or so, like myself, who made their way to the United Nations Rio+20 Summit.

PE: Well, of course, I've been in this game a long time, and I've always argued, as I just did, that personally, I'd like to have a planet where a billion or a billion and a half people can live the way you and I live more or less in perpetuity. I believe that will give us more *Homo sapiens* and *Homo sapiens* utility, in the long run, than seeing if you can cram fifteen billion people onto the planet living like battery chickens (those birds again) and have the whole system disintegrate.

MT: By "utility" you mean?

PE: The satisfaction one gets from using or consuming a product or service. I think we judge our utilities badly, but it's a personal choice shaped by the norms of our societies. I don't care. I use a lot of gadgets—we're both using gadgets—but I don't feel it gives me anywhere near the utility I could obtain in other circumstances. I'd be much more satisfied if everybody (including me) was working a thirty-hour week and having lots of time to read, make love, eat decent food, but not consume excessively—you know, more and more junk, or four Hummers in your garage, that sort of thing. Give me four wonderful lovers any time.

MT: And I make jokes very glibly about fast food being the food that should kill us as fast as possible—those who eat it.

PE: Which in my opinion it is designed to do, since it's designed to make people overeat dangerous food. But controlling fast food and the resultant type 2 diabetes, heart disease, and other plagues it contributes to will prove very difficult—since the whole process contributes to that ultimate value, growth in the economy. Look at how we allow cigarettes to be manufactured, even though the manufacturers murdered about a hundred million people in the last century and will murder more this century. But it's not clear how many lives could be spared by a ban.

WHAT ABOUT SOCIAL JUSTICE?

MT: Suppose junk food were legally banned. Would that save Medicare or Social Security from going bankrupt in 2034, or whenever? It would certainly put an end to this increasingly high premium we all pay for those who are obese, who are conditioned to smoking, or in other ways abusing their health. We are paying for them to be kept alive, but we're also supporting part of a huge medical establishment and the

A traditionally tattooed Inuit woman on Southampton Island in the Eastern Arctic, 1952.
© P. R. Ehrlich

rich lifestyles of executives in insurance companies, private hospitals, and Big Pharma. There's also the whole social justice side of it. The fact that in so many parts of the United States, economically marginalized families simply don't have access to nutritious inexpensive food. Or can only afford to live close to a freeway, their children taking in all of those pollutants at something like seven times the respiratory rate of their parents. Or whose main pleasure is from the nicotine that they draw into their lungs. Or in India children being spoiled senseless with junk food, high-carb hamburgers, pizza, and candy, with the resulting diabetes epidemic amongst the young. So you add poverty, stress, disease, junk food that derives mostly from animals that are tortured and slaughtered, with resulting widespread human obesity, and—wham! Welcome to the new world of ecological insanity; insanity that has been codified in so many ways into state and/or federal policies.

PE: I was on a Stanford alumni expedition not quite a year ago in Micronesia, mostly bird-watching, and was very depressed by the state of the coral reefs whose fish populations had declined dramatically since I last dived in the tropical Pacific some thirty years before. But one of the unexpectedly depressing things was that our driver guide stopped at a bodega to buy something. We went in; it was quite small and packed with shelves. There was nothing I could find in that place that I would eat. There were some five different kinds of breakfast cereal, each one 90 percent sugar, and that sort of thing. There were simply no healthy choices for the locals, for anybody.

A SCREAMING CABBAGE

MT: You asked me an hour ago, why are you vegetarian, and I tried to suggest that perhaps it's because (1) I'm able to be; (2) I'm disposed toward it because I have this Dr. Dolittle-like infatuation with everything I see that's alive, and for that matter that's dead. I love reading in cemeteries, I love touching rock—I know some monks who will argue that rock is alive, as much as I would argue that these plants are alive.

PE: You think plants are alive?

MT: Well, I've heard it said—I can't prove it . . .

PE: The thing I hate about vegetarians is they're not put off by the screaming of cabbage.

MT: I thought you might say that. And I could retort with all kinds of fancy philosophical and spiritual literature—and a copious literature, by the way—to back it up within various ethical traditions, principally Jain, which talk of five senses, or some of the Brahmanical Indian traditions that speak to the interdependency of all living creatures and, again, *ahimsa*, or non-violence. And it goes back to what you suggested earlier, namely, your identification of yourself or your wish to be thought of, or to think of yourself, as a practical idealist. Because anything short of that is anything but practical, given the state of affairs. But the screaming, or better, hollering cabbage aside—many are great, unsung tenors, you know—I suspect we are in 100 percent agreement on what you mean, both literally and subjectively, because I'm desperately concerned about the future of places like this beautiful mountain valley, the ability of our descendants to be able to have a walk and a few days of idle chatter, as we are doing, a step out of time in the most idealistic of places, for the most practical of reasons—what we've described as animal rights and wrongs. Cabbages and kings may have different perspectives, but I think as a species, we need to get our priorities sorted out, and quickly.

PE: I don't think this should be thought of as just a discussion of animal rights. Rather, we should use that topic as an example of how we have somewhat different views. For instance, if you had, as your idealistic goal, the maximization of the number of people who had decent diets on the planet, then you would have to eat some meat, because there are places you can "grow" meat (like grazing lands) and some places where you can't grow crops (like coral reefs).

THE GEOGRAPHY OF FOOD

MT: Well, there aren't too many places where you can feed animals whatever fodder you're going to give them, whatever

Stockyard. © Dancing Star Foundation

food type, short of industrialized agriculture, a petroleum-based economy that automates the growing, slaughtering, freezing, packaging, and distribution of those slabs of meat that once were living beings who would have grazed, in a true state of nature, uninterrupted by our human volition and incoherent taste buds.

PE: I'm speaking of just letting them graze. That avoids the industrialization part of it and most of the abuses those with an animal rights orientation rightly condemn.

MT: No, but if you can let them graze, then theoretically you can grow plants. As well as artificially growing meat, as you referenced earlier.

PE: No. Not the kind of crops to sustain a large human population, or at least not at the moment. And you also can't grow anything much edible in the way of plants on a coral reef or, in fact, in the oceans generally.

MT: You've got seaweed, for starters.

PE: You've got seaweed, but there's nothing like the quantity or the quality of food that you get in fish, which one could harvest sustainably. You can't survive on seaweed, and a huge number of people get a critical protein portion of their diet from artisanal fisheries, especially on coral reefs.

MT: But I take vitamin B12 pills. More than enough. So, what's your thought about twenty-two world fisheries in deep crisis, and so many individual threatened fish species, like the Patagonian toothfish or the Bocaccio rockfish, the Atlantic halibut and Acadian redfish?[1]

PE: Well, that's an interesting issue. Should we, in fact, be transplanting so much energy into farming a few kinds of fish to eat, which I think has generally proved to be an ecological disaster? No. I think the ideal thing would be to continue to harvest wild fish sustainably, if you had a small-enough human population. They provide critical protein to a lot of people, where small fishes often are just used as garnish, on rice usually. But the ethical position from my point of view would be to eat the animals if it could be done sustainably rather than let the people starve.

Bottom line, I think ethics is strictly a human invention, not a set of rules handed down by a supernatural creature use-

---

1. "Top Ten Most Endangered Fish Species," *Animal Planet*, http://animal.discovery.com/fish /fishing/top-10-most-endangered-fish.htm.

lessly postulated to have created the universe. (I say "uselessly" because all it does is raise the issue of the super-supercreature who created the supercreature—the famous infinite regress, turtles all the way down.) But ethics was a nice invention.

PROBABILITY DISTRIBUTION

PE: I could just consume anything I want and not say a word about what's going on. I am old enough not to care. I have enough money to eat all the meat I want to eat. I look at it as a probability distribution issue, namely, that the odds of my losing the lifestyle that I value before I die are considerably smaller than the odds are for you. And yours are considerably smaller than for my nineteen-year-old grandchild and enormously smaller than for my two-year-old great-grandchild.

MT: But if you really go that route—and I can't believe you do—at least not in day-to-day real life, then statistically just plot the probability factors of three generations of humans given all of the colliding indicators that are negative with respect to consumption, environmental degradation, loss of food stocks, loss of protein due to overfishing, overhunting, illegal trade in wildlife, destruction of habitat: it doesn't look good—the intensity of our consumption at every level. Hell, you are the one who first wrote the IPAT equation. It's even made it to *Wikipedia*, and I quote, "Human Impact (I) on the environment equals the product of P = Population, A = Affluence, T = Technology. This describes how our growing population, affluence, and technology contribute toward our environmental impact."

PE: I wouldn't be doing any of this if I didn't think there was still a chance. But that chance, in my view, is almost vanishingly small. I was asked recently, "What is your evaluation of the overall situation?" And I said, "Well, basically I feel we've maybe got a 10 percent chance of keeping civilization together, and if I work really hard, I might make it an 11 percent chance, and I think that's worth doing." And this colleague, a very distinguished scientist who has done great research on energy and economics, Jim Brown, said, "I agree with you, except you've got the decimal point in the wrong place." Jim says there is maybe a 1 percent chance of keeping civilization together, but it's worth the effort to try to make it 1.1 percent. You've got to remember: we need to revamp the entire energy-handling system of the planet in basically maybe a decade or so. And we haven't really shown much sign of even being interested in starting. But Jim and I (and you and Jane and Anne) all share the ethical decision that we must keep trying rather than just giving up and enjoying the life we could.

MT: So it's not going to happen.

PE: We're also going to have to change our water-handling infrastructure continuously for the next at least thousand years, and we haven't even got our present infrastructure in decent shape, to say nothing about rebuilding it for flexibility and continuous modification.

MT: In places like the Persian Gulf, there is more and more data suggesting rampant medical fallout from chemicals involved in the process of desalination. So you've got a new generation of medical researchers at places like King Saud University trying to understand the causes of CRF, chronic renal failure. For the moment, the victims as usual are rats, whose kidneys resemble those of humans.

PE: Michael, as we both know only too well, the biodiversity that runs our life-support systems is disappearing. We still have nuclear weapons pointed at Russia, and vice versa, for totally insane reasons. I mean there's not the slightest justification. There's no sign of the richest giant country doing anything about its population problem. Here we are in a nation where nobody's ever come up with even a semi-sane reason for having more than a 130 million or so Americans alive at once, and we are way past 300 million. And most of the economists and corporate leadership in this nation remain in favor of it growing forever, which is what the present trajectory suggests.

MT: I think the U.S. Census Report was last projecting something close to a half a billion Americans by century's end.

PE: And that's not a stopping point.

MT: No, in fact that suggests the bottom of a pyramid, potentially. The five-part piece on the "human deluge" and "Beyond 7 Billion" in the *Los Angeles Times* that kicked off July 22, 2012, referred to all the updated predictions, but only through 2050, at eleven billion, if the current fertility rates remain constant. They spoke of "the biggest generation in history," "youth bulges and political upheaval," the "age-old nemesis endures," China's one-child policy, and concluded with what they titled "Contraception and the Gospel of Life." They referenced you, Malthus, Rajasthan, Kenya, women in Nigeria, all the usual suspects. Pages A14–15 were titled "Barreling toward 11 Billion People." Finally, just days after the horrifying massacre in Aurora, Colorado, the *LA Times* had the guts to take on the population explosion front, right, and center. But they didn't mention that we will likely hit 10.9 billion by 2100 and could be 16.6 billion or even more. What the well-versed and clearly committed and unbiased author Kenneth R. Weiss, reporting from Jaipur, India, did say, was: "The global birthrate has been falling, but with billions of people in their fertile years, the population explosion is far from over." You and I—you for much longer—have been poring through the data worldwide for decades. The only thing different about this five-part piece is that it is on the cover of the *Los Angeles Times*. But nobody in Rajasthan—or Los Angeles, for that matter—is likely to change his/her reproductive choices as a result of the *Los Angeles Times*, in my opinion. Maybe I'm overly cynical.

SOMEDAY . . . CHANGE COULD COME

PE: And so it's very hard to be optimistic. On the other hand, I could argue that we have seen dramatic and sudden social change, even in our lifetimes. We've changed the racial situation in the United States; not enough, but dramatically. We've changed the role of women; not enough, but dramatically. When I was a kid, lynching was common in the South, and women's career choices were basically to be a nurse, secretary, telephone operator, or a schoolteacher. We watched the Soviet Union disintegrate when nobody expected it. So it's conceivable that some day in the not-too-distant future the world will wake up to its peril. Maybe it will grant equal rights and opportunities to women everywhere and provide every sexually active person with modern contraception and backup abortion. That way society could start to solve the population problem. It's just that I don't see much sign of it, since, for example, in the United States many state legislatures are now struggling to deprive women of control over their own bodies. And we're still in a period of "reproductive colonialism," in which the U.S. government is exporting American sexism and reproductive hang-ups. That's primarily in the form of the 1973 Helms Amendment that forbids U.S. aid recipients to fund abortion services and the amendment's provisions relative to rape and incest that are badly implemented. And there continue to be official attempts to restrict access to contraception, especially for sexually active young women. So I'm not counting on the United States climbing out of the reproductive Middle Ages soon and leading the world in trying to solve the population problem.

MT: Yet you still feel the advantage to your contributing that possible 1 percent of chance of change.

PE: I view it to be my duty, because that's the way my mother raised me. I just can't learn to face possible existential crises with equanimity.

MT: That's a good reason.

PE: And I don't have to make vast sacrifices in my own life to do it.

MT: You're honest about it. Most people are not.

BUTTERFLIES

MT: Tell me about the butterflies. You've studied them for most of your career.

PE: That one, an *Erebia*, belongs to an arctic alpine genus that's also found in Central Asia and Europe. There are a number of species of the same genus here in Colorado. Many of them are

Swallowtail butterfly (*Papilio* near *memnon*) in Malaysia. © M. C. Tobias

alpine. They presumably dispersed all the way down from the arctic tundra in the montane tundra—at high elevation—or retreated with the tundra ecosystems as the great glaciers retreated. I got fascinated with them when I was a kid, studied them in the Arctic, and did some work on their taxonomy.

MT: Why butterflies? What was it that motivated you to study them so thoroughly?

PE: Well, there is a combination of things. One is aesthetics. I was introduced to them in a summer camp when I was eleven or twelve, started a butterfly collection, fascinated by their variation, and got very interested in their taxonomy. I did my doctoral dissertation on what I started out thinking was their phylogenetic history—how they evolved and how different kinds of butterflies were related to each other and to other moths (butterflies are just a group of moths that managed to become active in daytime by evolving chemical defenses that allowed them to survive daytime predators like birds). The thesis turned out basically to be a taxonomic comparison using techniques that eventually evolved into so-called numerical taxonomy.

But I've always been a collector. I collect birds by twitching (trying to see bird species I've never seen before); I collect books compulsively. I collected butterflies by catching them and pinning them or by raising them from caterpillars, and was always thrilled by the variation displayed on their beautiful wings. But I'm color-blind. Had to depend on Anne to help me describe them.

MT: Red-green deficient?

PE: Red-green, standard Daltonism; some 8 percent of Caucasian males have it (shows how insane society's choice of traffic light colors is). So I did something that had hardly been done at all before. For my dissertation, I looked at butterflies worldwide, but the colors don't represent their relationships very well; structures do. The way you get the structure is to take the wings off and boil the body in potassium hydroxide and get rid of their covering of scales and their muscles and other soft parts. Then you can see and describe their skeletons in great detail. That is what I did for my doctoral research and came up with a classification that's still largely used today. That was more than fifty years ago. And then I started asking theoretical questions.

EVOLUTIONARY QUESTIONS

PE: Would the classification of butterflies based on their skeletons be the same as one based on their muscles? I did that for a while, but meanwhile I started asking evolutionary questions, and it turns out that butterflies in the field are ideal test organisms to ask evolutionary questions and ecological questions. They're so beloved by collectors that there is a huge body of information on their distributions and life histories gathered by amateurs.

When I got to Stanford, I wanted to do research on some important questions that have only been partially answered to this day in population biology. What causes populations to expand, to shrink, and what happens to their genetics when they expand and shrink? I started looking for a butterfly test system to use, since I knew they were ideal for such work—being almost as well-known as birds but much more readily studied and manipulated in the field. An amateur lepidopterist told me there was a population of checkerspot butterflies, *Euphydryas editha*, in a natural area on Stanford campus known as Jasper Ridge. In 1960 I started developing the system, beginning with what is known as a mark/release recapture study.

By giving each individual I caught a coded number with one of the early marking pens, Magic Markers, and then seeing where and when I recaptured them, I was able to determine how they moved around and how large their populations were. The very first year we began to make discoveries—that, for instance, there was not one population on Jasper Ridge but three. In a few years, we had learned a great deal about population dynamics, among other things that populations went extinct and would become reestablished by individuals immigrating from other populations.

It was early in the now more than half century that we'd been working on the system that it led to undoubtedly the most cited, theoretically interesting research project that I ever did with butterflies. I did it in conjunction with my friend, the brilliant botanical evolutionist Peter Raven. He was a colleague at Stanford, and we had a coffee break together each morning. One day I said to Peter, "You know, it's interesting the checkerspot butterflies that I'm working on feed on [plants of the families] Plantaginaceae and Scrophulariaceae, a weird combination." Peter said, "No, that's not weird." And he went into an explanation about how closely the plants were related, a whole raft of botanical stuff I didn't know at the time. I'd have fired any of my graduate students that carried that kind of botanical trivia around in his or her head.

And so we started talking about what butterflies ate, and gradually it became clear to us, first of all, that we knew much more about the foods of butterflies, say six thousand of some fifteen thousand species, than of any other large group of herbivores. It was incredible. The reason we knew those food choices is that people who collect butterflies want perfect specimens, and so they go out and find the larvae, the caterpillars (the eating-growing stage), raise them, and get the pupa (the stage in which an insect with "complete metamorphosis" is reorganized into an adult—the adult is the reproducing-dispersing stage). Then, when the butterfly emerges, as soon as its wings expand and dry, the collector kills it and mounts it for her collection. Then she writes a little note to an entomology journal describing what the caterpillar fed on. So Peter and I soon realized that we had a huge data set that we could glean from the literature.

Then it became crystal-clear to us that what all these different groups of butterflies fed on depended upon the secondary chemicals in the plants. The plants produced alkaloids or tannins, or whatever, and it then became apparent that those chemicals, instead of being excretory compounds, as the botanists then thought, were, in fact, defensive chemicals. It was ridiculous, in retrospect, to think that any organism would assemble a high-energy "excretory" compound and then, instead of excreting it, do the equivalent of stuffing it in its ear. Defending themselves was what the plants were doing. They can't run away from their predators, but they sure can poison them or blow their minds (we use many of their defensive compounds as medicines, and also as recreational drugs—like opium and pot). Plants are loaded with such stuff, and the realization of their role led us to the idea that butterflies and plants were in an evolutionary battle. The plants struggle to evolve nastier and nastier poisons to keep the butterflies off, and the insects are attempting evolutionarily to find ways of detoxifying the chemicals.

And when a butterfly solved the problem of getting rid of the chemicals that, say, make milkweeds poisonous, they "radiated"—that is evolved into many new species—feeding on the various milkweeds and were very successful. That's basically the story of the famous monarch butterfly and its many relatives around the world. And so Peter and I pointed out that similar things happened with predators and prey, with mutualists (species who help each other), with parasites and hosts, and so on. We published a paper on what we called "coevolution" in 1965. And that's been followed now by maybe fifty books on coevolution, or maybe a hundred, not to mention thousands of papers; a whole area of investigation, and it came about purely by dumb luck. Peter and I generated the data without ever looking at a living organism. We did it purely from the literature.

MT: So you're guesstimating there are something like fifteen thousand species of butterflies in the world?

PE: Roughly.

ONE IN A HUNDRED MILLION?

MT: And more to discover, presumably. I gather from the work I did filming our movie *Hotspots* with Russell Mitter-meier that, forgetting microbial organisms, we're looking at a hundred million species in the world potentially.

PE: If you count microorganisms, that's possible. There's a huge controversy on this, and it's really hard to say. It depends more than anything else on insects and nematodes, and one of the issues is how good is the taxonomy on beetles and so on. It also depends on how one defines "species"—a topic long ago settled, but still pursued by some who don't know there is no way to produce a universal definition (like trying to define "mountain"). Hotspots are a phenomenon, but there's much too much concentration on species hotspots, and there's much too much concentration on species extinction, especially when the big problem is *population* extinctions. A large diversity of populations is required to provide humanity with vital ecosystem services.

The problem is that many of the people involved in taxonomy have never had any real training in science and don't understand the need to sample the living world to understand it, rather than trying to describe it in its entirety. All you have to do is think about it for one second. Life is evolving rapidly and organisms going extinct so fast that even if you could describe everything, it's all going to be changed by the time you finish. Indeed, today much of it is likely to be gone in less than a century or so.

ON ETHICS AND BUTTERFLIES

PE: One of the troubles is that there are far too limited funds going into trying to save our life-support systems. It's a big

Professor Carol Boggs marking an individual of Gillette's checkerspot (*Euphydryas gillettii*) near the Rocky Mountain Biological Laboratory. Marking, releasing, and recapturing butterflies tells us much about their population sizes and movements. © P. R. Ehrlich

allocation issue, how much to spend on description and cataloguing and protecting species, as opposed to focusing on populations and the ecosystem services they supply. That's what much of the controversy is about. Is it more important, for instance, to maintain pest control services in the grain baskets of the world or protect narrow endemic species in tropical hotspots? Not an easy question to answer, and one with ethical dimensions.

MT: In all of your work—with butterflies, the evolution of coevolution as an idea, and as an observation—were there ever, in your early years working with butterflies, ethical discussions? Did it ever touch upon ethics?

PE: Well, it did for me because we discussed a lot of ethical issues when I was an undergraduate right after the Second

World War having to do with whether or not it was the right thing to spray DDT over all of New Jersey and kill off the butterflies and make the mosquitoes resistant to DDT. Building huge Levittown-type developments—named after the famed community in Nassau County, New York, with more than 17,000 homes—right on top of valuable habitat. Was it, for example, ethical to destroy natural habitat on a planet which has essentially none left—semi-natural habitat let's say now, in order to give people separate houses? Or would it be better to house them in decent high-rise buildings in already destroyed areas?

MT: I don't know. Allocation of land use is inherently biased. You have to ask: Who is making the call? Which constituencies have the most power in that call? Are we talking in biological terms or popular terms? Good governance, transparency, or profit? And whose profit? If we're speaking in strictly biological terms, who is actually on the deciding board? The committee that signs off on a decision? Is it in the hands of a courtroom, an arbitration hearing? City supervisors? Realtors? Lobbyists? Are we talking hotspots, cold spots, endemic species versus whole populations, populations of other species versus human populations, human populations versus landscape-size habitat?

All of this invokes a dizzying set of compounded debates within local, state, federal governments. The difficulties are augmented internationally, between nations, as was so evident at the Rio+20 United Nations Summit, whose best shot at consensus was broadly and non-bindingly confined to a lean document, "The Future We Want."[2]

---

2. "The Future We Want," Rio+20 United Nations Conference on Sustainable Development, http://www.uncsd2012.org/thefuturewewant.html.

One could argue that as these complexities increase, the solutions, or supposed solutions, are absolutely watered down until they have no traction whatsoever. In the case of the U.S. Fish and Wildlife, which I happen to be a big fan of, it's more and more focusing upon multi-tiered species protections as opposed to one-offs, which makes sense: it's called amortizing.

PE: But I think that's what's wrong with the Endangered Species Act of 1973, as I've already suggested.

MT: What do you mean?

PE: It's what I like to call "Darwin's mistake." I'm a great fan of Darwin. He was possibly the most brilliant human being who ever existed. He was much wiser than Newton (he had, unlike Newton, no fascination with the occult wisdom of the ancients and gave up religion early on) and had a much bigger impact on our views of humanity and ethics. But I wish he had called his groundbreaking book "The Differentiation of Populations," not *The Origin of Species*. This title and its implications have led people to focus much too much on the species-level of differentiation. That is a problem.

MT: But let's assume for a minute that the whole notion of species, as first formally recognized by Linnaeus botanically and then for fauna, by 1758 in his tenth edition of *Systema Naturae*, that all such nomenclature is peppered with real dangling modifiers that are so serious as to put in jeopardy the whole notion or validity of species demarcations.

PE: It's fine for one level of description—

MT: You mean species?

PE: Yes. But arguing about how to define "species" would be like geologists arguing about how to define "mountain." It's a perfectly usable and usually understandable term that defies close definition. It's similar to the problems of defining "culture" in anthropology, or "religion" in philosophy, or "tall" in human stature. "Kind" is probably the best simple synonym of species.

MT: OK. This is interesting. Perhaps it's the same situation with "ethics." Think about it: if ethics is too narrowly defined, then it excludes too many situations. But then, to hone in effectively on a specific example requires just that: specifics.

Just as "species" may be no less problematic than "mountain" in terms of really meaning something, so, too, "ethics" slides toward meaninglessness unless we are capable of using our words more instructively, more precisely. The worry with "ethics" is that it becomes preachy, religious, cultish. If anything, the twenty-first century has real problems that require real solutions. If ethics are to play a significant role in contributing to such solutions, they need an edge, which I take to mean realism. Ethics have to be real. To respond to real needs in ways befitting the best in humanity.

With that as a starting point, I would argue—as many have—that the real study of ethics requires an understanding and embrace of what is usually described as "normative ethics," or the study of ethical action. Ethical action connotes moral realism, in other words. Although you might argue that what is morally real for one person is not real for another. But regardless of differentiation and diversity, I think the only appropriate field of discussion in terms of ethics as it applies to human behavior and to biology would be ethics that is given to real action. Otherwise, we fall into the trap of the endless belaboring of points that lead nowhere; classical theories that drive students mad because they seem so utterly, hopelessly academic. Science, good science, has an implicit edge.

WHAT REALLY MATTERS?

MT: You yourself once said it to me: "If you could prove your hero, Darwin, wrong, you'd be the first to do so. Because that is good science." Similarly, good ethics requires not just having a context, but rallying behind a good cause. This is what we've been talking about for two days. What really matters at this point in history—with over seven billion human consumers and a crisis of extinctions, climate change, any number of other global debacles—is what we intend to do about it, in addition to simply discussing it. Our environmental issues have never had so profound an opportunity, in my opinion, to elicit huge, courageous, ethical convictions. Crisis has a way of getting one's attention. When the sky is truly falling, people tend to wake up. And that wake-up call is ethical to the core. People, I believe, prefer good to evil; clean water over filthy water; breathable air and healthy environments for their children. If we can agree on that principle of the positive as a universal, then I think we can also agree that as a species we are likely to embrace that which we can relate to. And that would be individuals who, like ourselves, are vulnerable, afraid, prone to pain. We're human, in other words.

If people will rally behind a cause that happens to be an individual—a girl trapped in a pipe in Texas, a kitten behind a brick wall in a New York deli, a ninety-pound adolescent cougar lost and cornered in a courtyard in downtown Santa Monica—I have to believe that they will rally behind any life-form if given a sufficient reason to do so. If firemen were climbing a tree and desperately trying to save a black bear cub, which they do all the time, and people who might not think twice about shooting one in season are riveted to their iPhones and iPads hoping to see that little bear cub rescued, then there is hope for the world. Maybe I'm being too simplistic or naive.

PE: Well, I certainly agree with the thrust of your argument—that's why I'm working so hard on the MAHB. I think we need a spiritual approach to the world, one that is, as I've said before, "quasi-religious." I want to see all the childish nonsense about supernatural entities who care deeply about our sex lives, who choose who suffers and dies when and how, who assign people to an afterlife where everyone who is interesting is tortured forever and where those too dumb to ever have had a little fun are doomed to float forever on a cloud listening to speeches by Norman Vincent Peale, disappear from human culture. I want it to be replaced by an evidence-based (not faith-based) society that cares deeply about how fairly its members are treated and how its life-support systems are treated. I want ethical questions to be on the top of the agenda, with everyone understanding the constraints set by scientific knowledge (no, you can't go to a heaven) and by the limits of science (no, evolution doesn't tell you if late-term abortion is ethical). And I think it's ethically required to do everything possible to move in that direction, in the hope that the science telling us it's likely too late is dead wrong. Does that fit in with why you're a vegetarian?

MT: Why am I a vegetarian? Perhaps, no, I can say definitively, it is because I'm deliberately living my life to the extent possible through the labyrinth of my extraordinary numbers of flaws and absolute hypocrisies . . . in a direction toward a better

declaration of myself; better behavior; more consistent appli-
cation of behavior on behalf of the convictions and principles
I espouse and believe in. There you see a worm wriggling in
the dirt. Now, here, at this moment, I rally around that worm
as I stand in its presence. I celebrate it. The thought of serving
it up on a fishhook to catch a fish is simply insane, immoral to
the extreme, a total and irrevocable insult to tens of millions
of years of cognitive, emotional and moral evolution.
I relate the harming of that worm unnecessarily to all
other forms of mindless harm. And that includes the unethical
choice of parents to have scores of children, in my opinion.

GETTING THE NARRATIVES RIGHT

PE: I would certainly subscribe to any campaign that aspires
to promulgate the appropriate messages for our time; the right
narratives that might hope to get us off this perpetual population and economic growth machine; this endlessly increasing
per capita consumption mind-set. As long as we're doing that,
as long as the population is growing and per capita consumption
among the rich is growing, we better take pause, all of us, because
there are no easy mythic or ideological fixes to make it all right.
On the worm, I'm less concerned—but then I love fresh trout,
even though paradoxically I don't like killing them personally.
Maybe that's why I prefer fishing with dynamite or rotenone.

MT: Eating them is every bit as much a piscicide as poisoning
them outright. But on another topic, what about tax incentives for people who choose not to have two kids?

PE: I think it's profoundly unethical in our country today to
have more than two children (unless the second pregnancy

leads to a multiple birth). Even possibly unethical to have two—1.5 might be an ideal average.

MT: That's the first time you've said it's unethical—to have more than two kids is unethical.

PE: In my opinion. And, again, two is marginal. If everybody just had two children maximum, then the political motto I would advocate would be to stop at two. Then, of course, your total fertility rate will go down below replacement, which is probably about where it ought to be, and where it's begun to move in the United States. But since many people might not limit themselves, some of us would be more ethical to have just one child or even none—and help reduce the odds that the present overshoot will catastrophically damage human life-support systems and wreck the lives of our grandchildren.

"EVERY CHILD A WANTED CHILD"

MT: So how politically or ethically do you respond to the couple who might enlist the slogan hailed by the UN for a whole decade, "Every child a wanted child," in order to theoretically justify, say, their having three beloved children? Especially, speaking theoretically, if one of them happens to be a famous scientist, another a major artist, and the third, an employer of a thousand needy employees?

PE: First of all you shouldn't have unwanted children. But you're being unethical if you want more than two because you're thinking of yourself, not about the world the children are going to inhabit. Yes, one or more might be famous artists or big-time employers, but also they could grow up to

Eastern Himalayan schoolchildren. © M. C. Tobias

be Hitlers, child-molesting priests, or ax murderers—or just more consumers feeding at a shrinking trough. If you love your children, you think of the kind of world you're leaving them and other children. The surest sign today that someone is unethical, or ignorant of how the world works, is if they have more than two kids. Things were different in the past, before the global situation became clear.

MT: I'm thinking of my mother's response and that of her generation, which is your generation. She'd say, "Wait a minute. How is it more selfish to want three than to want two?"

PE: The operative word is "want." Why is it more unethical for me to want to own five cars instead of one? One has to think about the consequences of one's actions, not just our own desires. But of course it is also clear that most people are unaware of the consequences of over-reproduction (or over-consumption), so they aren't breaking their own ethical standards.

In today's societies, it is the relatively rare individual who has acquired broad knowledge of how that society is maintained and works. It wasn't always so. When I lived with the Inuit, every member of the community essentially had possession of the entire Inuit culture, the group's body of non-genetic information. It was available to everybody. Today you and I are moderately educated people. But in terms of the amount of non-genetic information humanity corporately possesses, let's just say that we clearly don't have even one-billionth of the non-genetic information possessed by our society, no matter how educated you or I may be. There's now a giant "culture gap" between societies' corporate knowledge and that of individuals. And this means that you can have presidential candidates and legislators who do not think human beings are disrupting the climate, who believe in supernatural beings, and who know next to nothing about anything really significant.

And it means that many people will be puzzled by a standard that says having three kids in unethical. It likely wouldn't happen if we had much smaller societies, and education would see to it that people at least would have a better grasp of the culture gap and recognize the need to bridge it in key areas of understanding about how the world works. Certain parts of it you and I are trying to fill in with this book.

4 GETTING ONE'S PRIORITIES RIGHT

YOU OWE IT TO YOURSELF

MT: So, feeling as strongly as you do about that point, if a young promising graduate student comes to Stanford and wants to study entomology with Dr. Ehrlich, does Dr. Ehrlich say it would be more effective, more ethical if you'd get into the population arena and go to Washington and knock on doors?

PE: What Dr. Ehrlich would say is, first of all, you owe yourself something, so part of your life should go into what you enjoy, whether it saves the world or not. The second thing I'd say is that if you're going to have any influence whatsoever, you want to establish your reputation in science, because if you're just knocking on doors in Washington as Joe Blow, it won't make much impact, so you will want to build a scientific reputation. And then it depends upon what sort of entomology you're considering. If you want to look at issues of pesticide resistance, for instance, you could have a very big impact. If you want to classify butterflies, you are unlikely to have any impact at all, but you could have a fun life and put your spare time into the MAHB or some equivalent effort. And, finally,

I'd point out that the trend in the biological sciences is not to orient to a particular group of organisms as much as to a set of questions. Thus some questions about evolution might best be answered by studying butterflies, others by studying bacteria that live in your gut.

THE INDIVIDUAL VS. THE MOB

MT: Several points come to my mind, here. In thinking about Inuit society, small cultural enclaves versus, say, megacities, regardless of size, of scale, we're still talking about one species; the causes and consequences of an individual versus the causes and consequences of the collective.

PE: The new literature on mobs suggests that, actually, much of mob behavior is more sensible than any individual's behavior.

MT: Interesting. It calls to my mind the iconic image that Elias Canetti wrote of in his masterful book, *Crowds and Power* (1960), namely, a revolution, narrow streets, of people shutting themselves behind closed doors while the mob raced down those alleyways with pitchforks and guns. The alternative is to immerse oneself, out of self-interest, in the mob itself, to disappear within the crowd, to not stand out, to go with the flow. And you see this amid many studies of immigrants to this country who don't want to speak out, who don't want to make noise, who just want to sort of disappear and assimilate. So, in the end, ethics is predicated upon individual behavior multiplied by seven billion–plus *Homo sapiens*. Is there any argument that we can fashion—even theoretically—to suggest that if one person does the right thing, it can become an infectious way of thinking that triggers an overnight sea change? Like the Berlin Wall coming down? Or the disintegration of the Soviet Union? I'm thinking in terms of what typically precedes quantifiable change, such as the various models employed by great organizations like the Center for Biological Diversity in Tucson, Arizona, with a barrage of legal actions to help slow the pace of disappearing species and its programs that directly link overpopulation to extinction.

PE: There have been some attempts to model such things; voter models, for example, that depend upon how many people around you have to have a view before you are likely to acquire it for yourself. The causes of the Berlin Wall coming down were complex and largely misunderstood by the general public, although plenty of historians have had a field day looking at it. Interestingly, much of it was traceable to a few individuals, including Gorbachev, and Americans and Ronald Reagan basically had little to do with it.

MT: They just came out with a map today of germs around the world. They are actually able to map the worldwide spread of germs.

PE: There are billions of kinds of so-called germs, and the mapping is only partial.

MT: And at least twelve hundred pathogens, not to mention synthetic life-forms.

PE: More than a few politicians come to mind. Although, as far as I know, there is no true synthetic life-form as of yet. But there may be soon.

THE ROLE OF ETHICS ON A PLANET WITH LIFE-FORMS

MT: In this maelstrom, does ethics have a role to play with respect to any of the most seemingly intractable problems facing the world? Each and every life-form?

PE: Absolutely. I think it has *the* major role, and the problem is how you get people thinking about it. In other words, almost all the huge questions we face are fundamentally ethical questions. Let's say we're successfully feeding three and a half billion people and maybe three and a half billion not so successfully. Do we care whether that three and a half billion contingent of hungry, malnourished individuals shrinks or increases? Is that an ethical question?

MT: Is it fair to even raise that point?

PE: Once you say "fair," you're talking about ethics. Is it fair to have a world in which we have, in this country now, a very determined program of "Hood Robin" redistribution? Of taking money from the poor and giving it to the rich; or should we reverse that? What should be done about the collection of social parasites called Wall Street? When they're not stealing and doing their jobs legally, they cause even more damage by promoting economic growth among the already wealthy. Should Wall Street be disbanded and its denizens (many of whom can do arithmetic) be directed into socially useful activities like teaching middle school math? These are all ethical issues.

MT: I happen to agree that ethics is a thematic necessity that is spiritual, practical, idealistic, but brutal in its manner of expression, in that you get so many victims who are inno-

Old Delhi, India. © M. C. Tobias

cent and who have no voice and who are not part of any decision-making process. A situation in which those people—not to mention all the billions of other individuals of other species—that all they can hope for is a meal that night. Then the survival drama begins all over again the next morning. You were quoted in *National Geographic*'s 2011 special cover issue entitled "Seven Billion," regarding some of your early observations in New Delhi.

PE: Yes, they were quoting my book *The Population Bomb*. It's usually quoted as being racist because I dared to describe the visceral impression of a densely packed, economically marginalized megacity long before other scientists were willing to talk about that. Few people knew that Ralph Barr and I had helped organize sit-ins to desegregate the restaurants of Lawrence, Kansas, when I was a grad student in the 1950s, and soon after that started to publish on what biological nonsense racism was.

MT: In 1968, when I stepped off a plane in New Delhi and had my little passport stamped out on the tarmac and watched the Hanuman gray langurs racing back and forth and the little Pepsi-Cola stand and a telephone booth and hundreds of taxi drivers vying for my couple of rupees—and the musty odor and the density of sensations as I went into Delhi, where I would subsequently actually have an office at one point in my life and come to know the country really well—I, too, felt overwhelmed by the sensation of so many people. But no more so than in rush-hour traffic in Los Angeles. But of course I knew what you had experienced. I, too, experienced it. And then there's the whole notion of participation in democracy. What does that mean when we're nine or ten billion people? Whether in India, or the United States, or anywhere?

PE: As we discussed earlier, the Founding Fathers were concerned about a representative democracy and what could happen if we grew past a certain number of citizens. We had a population of about four million people in the newly United States in those days.

ARISTOTLE GOT IT RIGHT

MT: And I believe Aristotle said that if you have more than five thousand people, you cannot possibly stand atop the Parthenon, look down, and identify every individual in your community, which translates into sheer political chaos. Such numbers, according to Aristotle, exceed one's capacity to relate to other humans. That is synonymous with overpopulation. Thomas Malthus, on the other hand, as you well know, kept revising his opinions as his critics grew in force.

PE: Well, he thought we should get rid of the poor, basically.

MT: Aid to Africa, in his mind, would be absolutely preposterous, because you're simply prolonging and protracting agony.

PE: Since we created the crisis of poverty around the world . . . in other words, prior to the invention of Western imperialism, most people could have been just sort of making out. Just like the hunter-gatherers described in Marshall Sahlins's *Stone Age Economics*, they did all right. According to that view, hunter-gatherers mostly had a lot of leisure time. Poverty is sort of a construct that came along when agriculture was invented. Many analysts think the transition to farming was promoted by overpopulation. Whatever agriculture's roots, its immediate effects were largely to make people harder-working and less well-nourished. You can think of poverty as something that evolved in the agricultural world over the past five to ten thousand years (less than a twentieth of the time modern *Homo sapiens* has existed). And, yes, I mean poverty's obviously been in existence as far back as ancient Rome and Egypt, Sumer, and before.

BUDDHA ALSO GOT IT RIGHT

MT: There's probably always been inequity. During the time of Buddha, there were an estimated thirty-five million people in India. Buddha spoke to such disparities.

PE: It was basically a relatively equitable world until the agricultural revolution. Because after the revolution, you could settle down and acquire wealth. Such surplus would have been hard to monopolize when people were wandering, when they were hunter-gatherers. When people eventually produced enough surplus food agriculturally so that a family could supply food to more than one family, that let people support soldiers and priests and other people specializing in something different from procuring food, and some accumulating power in the process. The switch from procuring to producing was a gigantic revolution.

MT: The history of inequity and disparities between socio-economic classes has been covered enormously on so many levels and exacerbated by movements of wealth and wealth exploitation and extraction. We know the history to varying degrees from continent to continent, even as recently as post-1959 with the Antarctic Treaty. There's been wrangling in DC over its provisions in terms of what can we get out of Antarctica, what can we get from under the polar ice. As with virtually every square inch of the planet, massive institutions, multinationals with legally ordained personhood, as some term it, sit like giant implacable bulwarks of consumption around the planet, voracious to the extreme, grazing on everything, limitless in their greed to convert life into money; biology into their own power; the voiceless into consumables—speaking of other species—and anything, anything at all, that can yield a profit at any and all expense. Our hunting-gathering is now corporate, not individual. A billionaire sits in front of his/her computer, or issues buy-and-sell demands into a tiny headset, or—emboldened and politically bellicose—pushes a button to trigger some reprehensible action that could affect count-less others; or the bank, recently exposed after a seven-year investigation, in Costa Rica, which for a 1 percent user's fee had become the international cyber-bank of choice for global criminals of every persuasion. Millions of illegal transactions at the tip of one's fingers. In other words, a new Wild West high-tech culture that is capable of eroding the world's gains toward sustainability, social justice, biodiversity conservation, and all governance and transparency, ethical standards, and civilized standard operation procedures even more heinously than is happening at the allegedly "normal" rate of deviation.

PE: There likely will be a lot of oil under the once-virgin Arctic and Antarctic ice, and we'll probably be stupid enough to extract and burn it.

MT: That, too.

THE GAME OF DICE

MT: Even in remote Greenland, among many of the 52,000 or so permanent residents, apparently there is the sense that climate change represents a great new opportunity; that this will be a huge economic boon in terms of organic vegetables in newly ice-free regions. And they are probably right. Global warming *will* provide new opportunities for some, while destroying many others.

Soon, as the polar bear comes very close to extinction, you can bet that poachers of polar bears will become more and more systematic, ruthless, as the market for polar bears, dead or alive, surges. If ethics is, indeed, a principle of human behavior and of a survivable future, for not just our species but those that cohabit Earth with us, things are going to get extremely complicated. They already are. Can ethics and profit coexist? Certainly. Do they coexist? Not well, in my opinion.

PE: And given that mismatch, one would have to conclude that we're the weird life-forms, the bulls in the china shop. The microbes are the dominant ones. The truly successful ones.

MT: Is ethics systemic to our species, or do we have reason to believe it is evident in other species, given all the observations that have been reported upon?

PE: When I went to Tanzania to observe chimpanzees, I was determined to be non-anthropomorphic in my interpretations. It lasted until a baby chimp was scared by a snake-like stick or something and leapt into his mother's arms, and the mother cuddled it and patted it on the head. But that again doesn't necessarily imply anything about ethics.

MT: Well, does it or doesn't it?

PE: Again, you have to ask, "What are ethics?" And if ethics are agreed-upon standards of behavior, then chimps have no way to agree upon them, because they can't communicate the way humans do.

MT: But we don't know that. Why would you allege such a thing?

PE: Well, we have a pretty good idea how much they can communicate. Some things, absolutely. But not complex ideas. They can communicate "calm down, there's no danger," that sort of thing. But they can't use counterfactuals. They can't discuss plans for the future. They can't communicate such complex ideas and questions as "Is our population now so large that poorer members of our group are not getting enough food, and therefore we should increase the use of condoms and as a matter of ethics make safe abortion legal and accessible?"

MT: What about the Weddell seal, as well as other species, that will voluntarily inhibit reproductive success on the basis of food constraints which are present, and which they clearly acknowledge?

PE: The word "voluntarily" doesn't fit there. Does a kudu "voluntarily" run away from a lion? Do you know if there is any reality outside your head?

MT: No. But there is, as you know, a predisposition in your head and in my head, along with seven billion other human heads—or minds, if you will—to posit a world outside ourselves; a world in which we live, participate, and depend entirely upon for our air, water, food, and so forth. A world that we have agreed to agree upon, forgetting nuances, degrees, perspectives. A world in which we would nearly all concur that there is at least *something* going on outside of ourselves. If we are all wrong, then so be it. We shall then have collectivized a grand delusion, which by itself holds enormous interest, if not at least a modicum of promise.

But, of course, in any such world there *are* nuances and differences which seven billion *Homo sapiens* are bound to debate. In this context, of ethics, of other species, of our decisions or foregone conclusions, there are particularly acute considerations. I am less impressed by the many scientists who have considered numerous examples from the encyclopedia of animal behaviors to be considered "moral" in their own right. There has been enough of a sea change in the ethological community—the breakthroughs by Goodall and others who do not hesitate to call animals by names or anthropomorphize in any number of ways. It's rather hard not to when, for example, you see mountain gorillas defusing snares that have been used by poachers to kill other mountain gorillas. Or any number of other incidents. But what especially impresses me are all the incidents of clear biophilia. That incredible instance, for example, of the giant turtle and the baby hippo that bonded on a lonely shoreline of Somalia, when they were stranded by the Christmas 2004 Indian Ocean tsunami that killed hundreds of thousands of people. Or cats falling in love, so to speak, with mice. Or that famous sequence of photographs in a *National Geographic* showing a polar bear playing, really playing, with a husky, the two of them clearly delighting in the companionship, before the polar bear eventually sauntered away. I believe it occurred in Churchill, Canada. There are just so many anecdotal images that capture the essence of true feelings. We know they are feelings, because we feel something when we see such pictures. And I have always said that the soul speaks when it is spoken to.

PE: I suspect these are all a combination of bad interpretation, extraordinary circumstances, and who knows what else. Virtually all of those interactions are open to explanations in which "essence of true feelings" wouldn't play any role. Straightforward ethic-less predation and "cruelty" is the constant, everyday activity that outweighs these examples of "animal ethics" billions to one. But one could also discover that our views are merely solipsistic, in which case we're all part of an unimaginable dead end.

MT: That's an interesting dead end, because it's a temptation to go in that direction, but it leads to flailing in the wind, because there's nowhere else to go, there's nowhere to take it. So does that version of fatalism lead to mere self-importance, selfless importance, to science, to ethics, to politics, to economics, to family, to anything at all? And what about God?

PE: I think you're on a sidetrack. We have a big-enough problem figuring out what should be, and is, or can be ethical amongst us human beings who communicate with one another. To ask whether or not tapeworms have ethics . . .

MT: Granted, that might not help us, or not in this generation.

PE: The only thing we share with tapeworms is our food and shit.

MT: We wriggle a lot. But if we are prepared to exclude worms, then what other species are we prepared to exclude? Chimpanzees? Elephants? Macaws? Dolphins?

PE: *All* of them when it comes to whether they are ethical (or moral). I'm speaking of responsibilities toward one another. I mean, after all, you and I would agree that we also have ethical responsibilities toward three-month-old infants, even though they don't have ethics themselves. That's why I'm against newborn males being tortured for the pleasure of old men. You can be a target of ethics without language; you just can't establish ethics or decide what's moral without language with syntax.

MT: That's strangely draconian. You're lying on a bed in Florida paralyzed, and one member of your family wants to euthanize you and the other doesn't, and the law chimes in— you may scream bloody hell, and we know that case, and we know other cases, and the controversies. So how are humans to debate ethics in a manner that will result in something meaningful? That really cuts to the chase and isn't garbled.

PE: Much of the debate comes before the paralysis, and in cases of people in comas, the debate also goes on around them.

But those chimpanzees at Gombe couldn't debate whether or not it was ethical to attack the female and kill and eat her youngster—or carry it back to Jane's porch. In any case, I am to a certain degree utilitarian, but I'm not going to be necessarily utilitarian in everything. I'm a practical ethicist, let's put it that way.

ENVIRONMENTAL HISTORY

MT: Well, when one looks back at all of the environmental battles in documented history—

PE: We lost almost all of them.

MT: Not all of them. The Lacey Act at the turn of the century, the Endangered Species Act, the EPA, Clean Air, Clean Water . . .

PE: All steps in the right direction that have fundamentally failed in their missions. Despite the Lacey Act and ESA, biodiversity is disappearing ever faster, dozens or more times faster than the "background" rate that has occurred in between the handful of huge past extinction episodes. Air is "cleaner" in some areas, but the persistent organic pollutants (POPs), endocrine disrupting compounds (EDCs), and CO_2 make it ever more dangerous. Less obvious garbage in water in some places, but the invisible replacements like antibiotics that are now replacing natural microfloras over large areas, and pole-to-pole contamination with EDCs pose a far greater threat than visible trash ever did. So likely do the micrograins of plastic that now invisibly pollute the oceans. We've lost because the factors wrecking our life-support systems are escalating much more rapidly than civilization's pathetic attempts to deal with them. China alone, trying to emulate America's mistakes, could bring down civilization.

MT: So if ethics predicates any hope of a solution for the dominant problems confronting our species, and all the species over which we are ungainly stewards and shepherds, if we even care at all, then what are the most critical watershed issues that everyone should be talking about? What should the headlines read?

WHICH HEADLINES DO YOU READ, WHICH DO YOU IGNORE?

PE: They should be about what the MAHB is doing. That is, facing up to the fact that telling people about the science of the human predicament and trying to point out the major drivers—overpopulation and overconsumption among the rich—just doesn't lead to sufficiently rapid behavioral change. Human cultures are just not evolving fast enough. We know that from both personal observation (consider the failure to really convince most people on climate disruption in the face of the successful campaign of outright lies by the paid climate "deniers" and crazed anti-regulation ideologues) and from social science research that simply informing people what the science says doesn't change their behavior.

And so the issue is what kind of narratives, what kind of approaches do you have to take to get people to change their behavior in the direction of sustainability?

There are a lot of examples, both great and small, of how it can be accomplished. In terms of saving energy, for example, it turns out from studies done about trying to get people to reuse hotel towels and persuading people in homes with smart meters to conserve energy, that the incentive for change is not really saving the environment or even saving money. Instead, it's what one's neighbors, or people who previously stayed in the same hotel room, may have done. If people find out that all their neighbors are saving energy by adopting a certain change of behavior, then change indeed will come. Robert Cialdini, the brilliant social scientist who did the research, refers to the pattern as reacting to "provincial norms"—basically, a geographically narrowly bounded ethical universe.

MT: That is something we all grew up with in kindergarten. It's called peer pressure.

PE: Actually, there is no real peer pressure—people react not to what they're told or observe, but to what they think their peers are doing. But again, when ecologists, for instance, want to convince people to save biodiversity, they say science tells us biodiversity does X, Y, and Z, but that message is not getting through. We're not even having the discussion. As you know, the vast majority of politicians and economists don't know what an ecosystem service is. Another failure—this one of our educational system and the corporate-controlled mainstream media; you'll never have a clue about how the world works from watching *FOX News*, listening to Rush Limbaugh, or reading the editorial pages of the *Wall Street Journal*.

MT: And we have a president who until the disaster at Sandy Hook elementary school refused to take on gun control; a Republican Party that is bought and paid for by the lobbying organizations of the gun nuts and utterly and hopelessly in denial on almost every environmental issue; and a general public that is fundamentally shy of, or outright hostile to, the notion of population control, in any form. It is nothing short of remarkable that, as I mentioned earlier, the *Los Angeles Times* ran its five-part piece on the topic.

SEX EDUCATION

PE: Sex and reproduction in our sex-soaked society is another area of social failure. There is little sound sex education anywhere, and many if not most people are confused about sexual issues or their own sexuality. It's not restricted to the uneducated. I once had a fellow graduate student who asked me if I wouldn't mind buying some condoms for him, and I said why don't you buy yourself some condoms? And he said he'd be much too embarrassed to go into a drugstore (this was in the 1950s) and ask for condoms. And I said, "Well, go out to the Dynamite," which was the local roadhouse, "go into the men's room, and you'll find a machine there. There's no person, you just put a quarter in the machine, you turn this little handle, and you have a condom." And he said that wouldn't be any good because those were "only for the prevention of disease." So I actually knew the only person in the universe who believed the politically correct lie that those condoms were only for the prevention of disease. People sadly often believe whatever they're told, even if they are told it by poor Rush Limbaugh. My friend was a smart science student, but his skepticism was mostly restricted to his thinking about science.

Think of the influence on sex of the big unethical monotheisms that run much of our Western world and believe themselves to be above criticism. Prior to their dominance, the pagans in the Roman Empire let each other live and let live— as long as you helped protect your fellow citizens by doing the proper sacrifices to the local deities. The pagans weren't interested in converting others. Of course, the most unethical thing going on now with one of the monotheisms, Catholicism, is opposition to the use of contraception. The main source of that is the Vatican and its bishops. Yet Catholics use contraception as much as non-Catholics, and they have abortions with even higher frequency. But the reason that the hierarchy fights against both is that the higher-ups in the church don't want to admit that the Protestants and Jews were right. The picture with Islam is more mixed and not remotely as injurious as the Catholic position. Islam, like Protestantism and originally henotheistic Judaism, generally permits contraception, and since the Pharisees, Judaism has had little interest in proselytizing. All are unethical since their leaders often attempt to dictate the behavior, especially sexual behavior, of others with at the very least threats and at the worst torture and murder. All of them, with trivial exceptions, tend to oppress women. Much of this is an evolutionary continuation of male dominance, based now often on dictates received from imaginary entities. And, of course, in most societies criticizing the ridiculous ideas of religion cannot be part of reasonable social discourse.

Thus you have "God-fearing" people trying to maintain their rigid positions, especially trying to control the lives of women. I consider that their rigid opposition to something so basic, so critical to the future of life on Earth, as controlling reproduction to be just as unethical as any major affront to the environment or terrorist act. They're working to kill people— women who need safe abortions now, and our descendants who are likely to have much higher death rates related to the decay of human life-support systems as a consequence of overpopulation. The pope and many of the bishops are one of the truly evil, regressive forces on the planet, in my opinion, interested primarily in maintaining their power. What other collective conclusions should one reasonably draw from the outrageous lawsuits that have been filed on behalf of so many victims of pedophilia and other crimes against humanity, all covered up by the church?

Consider the bishops assaulting nuns for trying to be good Christians and helping people, while turning their backs on the child abuse that has become one of the defining attributes of the church. And it's not just the Catholic Church. Think of the state of women in general.

MT: I remember a story from my research in Nairobi. A member of the Vatican who had been visiting some of the dioceses in Kenya went into a Catholic high school—and totally at random, one of the school authorities opened up a student's locker, and there they found some condoms. Ten years of good works by numerous family-planning NGOs and the United Nations Population Fund, the UNFPA, were finished in one Vatican-driven mindless instant. I suppose what troubles me the most about such things is the challenge that survives: What are the most effective arguments for changing human behavior? The finest models we can imagine for implementing wise land use, animal protection, for social justice at all levels?

PE: All you have to do is turn on the TV to get depressed.

CULTIVATE YOUR GARDEN

MT: There are so many priorities and so many wildfires erupting simultaneously around the planet. Two weeks prior to the Rio+20 Summit, nation-states, one by one, waffled over whether to even go or not. It was chaos, evoking the poetry of John Donne, and the way Ernest Hemmingway quoted him

by way of a prefatory remark in *For Whom the Bell Tolls*. When
you wake up in the morning, do you go back to sleep, or do
you choose to remain awake? Because the challenges are so
numerous, so daunting. I had this very depressing discussion
with my mother last night, who said you just have to smile and
be happy and be grateful for what you've got. Because I sort of
gave her a summary of what you and I had discussed yesterday,
before the mosquitoes drove us away, and she said, "Oh my
God, it's so depressing." And then she pointed out that everybody is aware of all this. That we don't need scientists to talk
about how bad things are in the world and the coming cata-
clysm. And I said, "Mother, it's not coming, it came, and it's
here, and it's continuing, and it's getting worse—it's escalating." She said she was too old to think about these things. That
she just wanted to spend the rest of her days with her friends,
reading good books, cultivating her garden. Which happens
to be a terrific garden, in Denver. Not a bad city, as cities go.

THE GALÁPAGOS

PE: First of all, we're struggling with the MAHB and through
it trying to mobilize what, in my opinion, is essentially an ecologically uneducated population. Which makes every problem
that much more difficult to attack, let alone resolve. I think
on the ethical front, you know, you can look at the Galápagos
as an example and realize that a biological and cultural site of
enormous importance to all peoples is under great threat. The
islands hold a fascinating and unique flora and fauna and are
a key site in one of the greatest intellectual advances of our
civilization—the uncovering of the mechanisms of evolution.

There were about 1,000 visitors a year in the 1960s. Now
there are some 100,000. The human resident population has

more than tripled, and many invasive species have been introduced. Overfishing is endemic. Yet the ecological ignorance,
overpopulation, and overconsumption that are the fundamental drivers of this ethically depressing situation can't be solved
just within the Galápagos or even within Ecuador. Poor people
need to make a living, and rich people don't see the big picture
or (unethically?) don't seem to care.

One can look at any number of problems—local, regional,
or global—and, in the end, issues in ethics are major road-
blocks to their resolution. Think of the culture of an Inuit
group that declares they have cultural rights to kill whales. But
at the same time, by killing those whales, that culture is also
killing global bio-heritage, "property" of all humanity. Moreover, those whales are going to disappear, and then the Inuit
won't have the cultural privilege of killing them. Then what?

People in the Galápagos would seem to have a right to live
by fishing; but when the fish are gone, what will they do?
One of the curses of an education in ecology is the frequent
need to ask the question, "Then what?" Some claim that it is
every American's "right" to have as many SUVs as he or she
can afford. But when climate disruption, in part due to SUV
exhausts, wrecks agriculture, and people are starving, then
what? Shouldn't SUVs be banned right now and most other
automobiles phased out over a few decades? Others say that
women have the "right" to have as many children as they want.
But if they want four apiece and, as a result, there's barely food
for one, then what?

MT: Like the Southern California fisherman who testified that
there should be a limit on the number of sea otters in order
to balance the rights of fishermen with sea otters. And yet the
boundaries and population numbers established by humans

out near San Nicholas Island are purely arbitrary. Now, the U.S. Fish and Wildlife Service is trying heroically to restore sea otter populations a hundred years after the mammals were driven to near extinction. Competing altruisms, typically favoring human demand.

THE NECESSITY OF HUMAN POPULATION CONTROL

PE: And if we don't solve the issues of population growth and consumption, all the rest of these issues won't stand a chance of being remedied. Whatever your cause, it's a lost cause without population control. Every day that is more obvious. So my basic view is that humanity ought to stick with the fundamentals. You are not going to solve the problems one at a time. The Grameen Bank, for example, is a good thing that's helped some people. But compared to what's going on in the world as a whole, it's like trying to bail out the ocean with a thimble. Much the same can be said about many examples of positive trends. All too often they are too little and already too late.

MT: But, of course, that whole concept of micro-financing— for which the Grameen Bank was awarded the Nobel Peace Prize, and particularly in cultures where women have been held back for so long, as in Muhammad Yunus's Bangladesh—is brilliant.

PE: No question about it. It's just—you know I'd be cheery if we had another thousand years instead of maybe a decade, if we're lucky. Many small-scale successes; nowhere near enough large-scale ones. That's why society needs rescaling—we've got to reduce the size of the entire human enterprise.

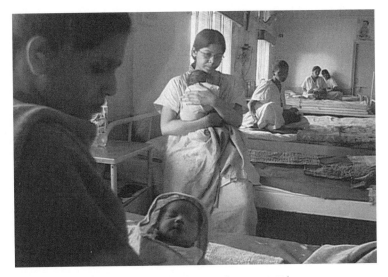

A superb family health clinic in India. © M. C. Tobias

MT: Well, if we only have a decade if we're lucky, why are we sitting here?

PE: One reason is that it's tough to know what to do. I'll tell you one thing: it's clear that Congress is doing absolutely nothing to solve the big problems. Politicians so far have been hopeless. The most overpopulated nation in the world had a presidential election in 2012 in which not a single serious issue was debated—and certainly not population control. The debt "crisis," for example, is completely solvable by negotiations among people. It would be tough and involve some redistribution of wealth, but all it would require is the will to do it. In contrast, you can't negotiate with nature. You can't say, "Nature, we're going to bust through the two-degree temperature rise 'safe' limit, so you'll have to let us grow enough food after a five-degree rise." Look at what *was* debated. That "gay rights" and the right of a woman to control her own reproduction are still debated shows that the United States is still in the grip of religious-based prejudice—that the ghosts of Dred Scott and Roger Taney still haunt us.

Another reason I just sit here now is that I believe ethically you owe something to yourself and your friends. Anne and I fight the good fight all year long, but being able to fight it from here for a month or so makes a big difference. Among other things, it reminds me of how much beauty remains in the "wild" world despite the efforts of Wall Street and the fossil fuel industries to destroy it.

MT: So structurally, the key issues from your perspective are population and consumption. If we don't get those two under control, we're finished. All right, if that's the case, then the consideration of ethics and the wide expression of it in so many avenues of thought and investigation are, in many ways, irrelevant?

ARE ETHICS OBSOLETE?

PE: Yes, sadly, much ethical discussion today is largely irrelevant, such as whether it's ethical to eat pork, be homosexual, or take an imaginary deity's name in vain—just the ethics of nonsense. But the big ethical issues—those about the life and death of billions of people and our civilization—simply overwhelm the minor ones, such as those just mentioned, or even whether it's ethical to use tax shelters. As Gandhi once said, "There are people in the world so hungry, that God cannot appear to them except in the form of bread." Naturally, most people in economic distress want more consumption. So that immediately raises one of the major ethical issues: How much is it incumbent on us rich people to share?

MT: Well, here's one perfect case example of that. I read years ago that Malaysia wanted to develop its own indigenous auto industry, and economists apparently recommended a pronatalist policy that would grow the population to a threshold of twenty-six million to provide sufficient consumerism for a home-grown automobile.

PE: And then they'd have to give it tariff protection.

MT: Absolutely. And the whole Southeast Asian coalition of economies would band together to ensure that there'd be new automobiles in India, which of course there now are. One of the smallest cars in the world and one of the least expensive is the new Indian Maruti 800, I believe. It's said to be excellent.

PE: The last time I was in Delhi, I was nearly run over by someone driving a Lexus. The point is, whether across Asia or Manhattan—Wall Street loves more consumption, but they also love keeping all the money to themselves. As much as I was no admirer of many of the things that he did and said, Henry Ford was fundamentally smart. He designed the car, prices, and the wages so that his employees could afford a car. That's smart. Whereas taking money from the poor and giving it to the rich sooner or later will destroy a consumer society. But ecologically speaking, it is no longer appropriate to be continuing to build a consumer society, since what we all want and what we're all doing is totally out of sync with sustaining a livable world. The totality does not mesh.

A CRITICAL MASS

MT: So if we're going to envision some form of positive and essential critical mass in time to save the world from ourselves, what are we going to do? What is it going to look like? How do we begin to wrap our minds and hearts around this global emergency? You've indicated in so many ways throughout your career that we must first get people to think about making the right choices when it comes to their fertility rates.

PE: The right choices for them and for their kids and grandkids.

MT: OK. Knowing as much as you do about the fickleness of human nature—or natures, as you've rightly put it—why have you devoted so much of your research to butterflies since . . . ?

PE: Nineteen forty-seven.

MT: And of all places, here. When there are some fifteen thousand or so butterfly and skipper species around the world. Why here?

PE: This region, the East River Valley, is one of the richest areas for butterflies in North America, with some seventy or so species.

MT: What are the preconditions ecologically that make this area so attractive for them?

WHY HERE?

PE: You have big mountains so, as I said, those butterflies that live in the tundra and the far Arctic can migrate at high elevation to this region. Other butterflies move up from the

Early spring at John Harte's global warming experiment at the Rocky Mountain Biological Laboratory. Control plots are between the partially melted experimental plots with the heaters hanging over them. Over the twenty-three years the experiment has been running, the unheated plots are detectably changing in response to real climate warming the same way, but at a slower pace, as the heated plots have responded to intense experimental warming. Photograph by Dr. Scott Saleska.

Southwest and even Mexico each summer and often take several generations in the process, but they get here, say, by the end of July. You have species coming from the eastern slope of the Rockies—some that can be found all the way to the eastern United States—or come from the Great Basin or from throughout the western United States. Basically four different butterfly biogeographic elements come together in this specific place which happens to be, at least so far, the wettest part of Colorado.

MT: So you've got precipitation—even now, as the southern half of North America endures its most sustained megadrought since the 1300s—probably the same scale of drought that forced the cliff dwellers at Mesa Verde, not all that far from here, to vanish. And clearly there is a very broad array of vascular plants here.

PE: And there's also great altitudinal diversity. Nearby Gunnison itself is a little over 7,000 feet, and the mountain that we're looking at now is over 12,000 feet high, so you get a 5,000-foot elevation gradient, at least in this valley.

When we take our hike later today with Dr. John Harte from Berkeley, it will be especially interesting and challenging to try to make the bridge between ethics-based behavior and thought, and the science of biology. As you know, Harte is a natural resource expert, physicist, biologist, energy guru, climate-change maverick, and advocate for waking up society. He's a researcher who has been studying here at the Rocky Mountain Biological Laboratory (RMBL) for many, many years. He's a very dear friend of mine. And he's an outstanding expert in terms of ethics, because considering the plight of humanity, he shifted his research from theoretical physics to ecology. He did his first big project in Florida in the 1970s on issues related to the Everglades. Later he came here and worked on acid rain, and then he shifted to working on catastrophic climate change. So he's let his career be guided by the needs of society. That, in my view, is highly ethical. Other scientists sometimes just continue doing what interests them, which is fine and in the long run could possibly help society. But only if we succeed in having a long run.

MT: OK. Your population studies, of course, made you a household name in 1968, much like an ironing board or *I Love Lucy* and *My Three Sons*. But did your studies of butterfly populations in any way translate into human population work?

BUTTERFLY DATA

PE: Yes. First of all, the same kind of mathematics is used in studying butterfly populations as studying human populations. But I got involved in so many things, in part, because of discussions I had with World War II veteran roommates when I was an undergraduate. At the same time, I was seeing the effects of overuse of DDT on butterfly populations in New Jersey, and I was watching habitats where I had collected butterflies being paved over as the human population expanded. So my interest in the issues that I've pursued for basically my entire life was in large part generated by the sudden explosion of consumption following the Second World War.

MT: Yesterday, when I asked you how you developed your passion for butterflies, the first thing you said was the aesthetics, the beauty, as you perceived it as a kid. So here you went from a child in love with butterflies because they're so pretty to someone hearing about the horrors of World War II and then seeing the very areas of beauty and scientific inquiry devastated by American overconsumption and rapacious development.

THE LESSONS OF WAR

PE: There are all kinds of lessons that come out of that. For example, we know from mathematics basically that it will take many decades to make big changes in the size of the population humanely. But of course the experience of World War II also taught me that if we want to change our consumption patterns, we can do it almost instantly. In 1941 the United States produced almost four million passenger cars. Then on December 7, 1941, the Japanese attacked Pearl Harbor, and by 1942 we weren't producing any automobiles—what we were producing were military trucks and tanks. We produced something like . . . I think it was 350,000 military aircraft. We produced hundreds of thousands of machine guns, howitzers, mortars. We did all of the work to develop and then deploy two different kinds of nuclear weapons, all within four years. We rationed gasoline, we rationed meat, we rationed sugar. I went out and collected milkweed pods because the little umbrellas the seeds float on are good stuffing for life jackets. And sadly, we managed somehow to keep society together, even though some 400,000 of our people in the military were killed.

Then, in 1945 we turned the whole thing around again, and by 1946–48 we were producing mobs of automobiles, TVs, Levittowns, and the like. So you know that with the right incentives, society can do the right things fast on the consumption side. It can't do the right thing fast on the population side, which is the reason humanity should have started fifty years ago dealing effectively with the population issue.

The United States, when I was a grade school student, had 130 million people. As I said earlier, no one has ever come up with an even semi-sane reason for having more than 130 million Americans living at once. And the only excuse for that many is that we fought the greatest land war in history with that many. Not that numbers mean that much militarily— the Germans had only about half as many, but had a much more effective army. Anyway, if you want to have a maximum number of Americans live, you want to engineer a sustainable population for millions of years, not try to see how many you can cram into our borders in a century or less until the United States collapses.

My experience of World War II taught me something else. I was nine to thirteen years old and paid close attention to the campaigns. I watched bloodless propaganda films like *Guadal-canal Diary* and *The Fighting Lady*. I was anxious for the war to last long enough for me to fight in it. It shows how stupid having unmyelinated frontal lobes can make a kid. Human brains don't have full capacity for judgment and social control until the mid-twenties. It's no coincidence that cannon fodder, mostly teenage boys and those in their early twenties, are called "infantry" (from the Latin for "infant").

. . .

PE: Jesus, did you hear that plane? Even in the mountains they are frequently overhead. I never realized how much noise pollution there was almost everywhere until, when I was a correspondent for *NBC News*, I started doing outdoor interviews. That, recording birdsongs, and trying to listen to taped lectures as I walk to work really convinced me how bad it is.

But I'm no Luddite. I'm an instrument-rated multi-engine pilot, and I'm sorry I had to give up flying for financial and environmental reasons. Airline travel is no longer fun. I've known Peter Raven for more than fifty years. I tried to get to his birthday party a month ago. Unfortunately, United Airlines canceled the flight while Anne and I were in the car heading to the airport! I wish I could have flown myself—just as I wish there were only a billion people or so in the world so that light aviation would be a reasonable form of transport in some circumstances. But as population grows in a world where we can now no longer afford to use fossil fuels, even commercial aviation is likely to go into a great decline in a few decades at most.

5 THE BIOLOGICAL FUTURE

CLIMATE CHANGE IN THE ROCKIES

Dr. John Harte. © M. C. Tobias

JOHN HARTE, *hereafter* **JH:** I became interested in climate change about twenty-five years ago. I became very interested in how global warming might affect ecosystems and in how changing ecosystems might affect the climate through what we call feedback. This project at warming meadow began in 1988; that's when we designed the system—and it was the first experiment of its kind—to heat an ecosystem at a level that's comparable to what we expected the world to be heated to in about fifty years.

We use electric heaters, radiators actually, to heat the plants and the soil; they are turned on day and night, summer and winter, and they've been on for over twenty years. There are ten plots, and half of them—the ones with the heaters above them—are the heated plots, and in between them, like this one here with no heater, these we call control plots. And each plot is ten feet by thirty feet, and the plots are, as you can see, vegetated. There are creatures living in the soil, of course. Pollinators come and pollinate the plants, obviously. And we've been studying how the heating is affecting these ecosystems.

Two degrees Celsius is the amount these heaters are warming the soil to a depth of about six inches. Now, at the surface, the heaters cause more heating, three or four degrees, and actually it is what we term a load level of heating. When we set up this experiment in 1988, the projection for global warming in the year 2050 was a two-degree warming. This is two degrees Celsius, not Fahrenheit. So in the Fahrenheit scale we're talking about four degrees. But now we know that global warming will be even more severe than that. By the year 2050, we expect that there may be as much as four or five degrees Celsius of heating. And so what we're doing is actually a slightly lower level of warmth than will probably occur by the year 2050. Nevertheless, even though we've underestimated the warming, we are seeing profoundly important effects showing up, even after just five or ten years of heating. In fact, the effects have been huge and cumulative. A small amount of heating, just a couple degrees, can cause such big effects.

Back in the late eighties, we had no idea what the future of our ecosystems might look like. In fact, people weren't even really asking the question. Most of the focus in global warming research was asking just how hot will the planet get. And I wanted to be able to anticipate what ecosystems would look like. I wanted to see global warming before it sees us, and that's why we set up this experiment. The heaters are causing snow to melt two or three weeks earlier in spring in the heated plots than in the control areas, and that will happen all over this area. It's increasingly happening. We're beginning to see a trend toward earlier snowmelt in the climate record from year to year. So what we're doing is greatly advancing snowmelt in the heated plots so that we can look forty years ahead, and the effects of this have been huge.

MT: What sorts of effects?

JH: One of the most important effects is that our heated plots have lost about 25 percent of their soil carbon. Carbon is stored in soil in the form of organic matter or humus, and when you heat these plots, the carbon in the soil burns off as CO_2, carbon dioxide. Read: global warming. So these heated plots are contributing to the warming effect. Now you might wonder, is 25 percent of soil carbon a lot? Well, let me put it in perspective. If you look around the world at all the Earth's soils, down to a depth of about a foot, what you find is approximately four times more carbon locked up in soil organic matter than is now in the atmosphere. So if you release a fourth of that, you get an amount equal to the amount in the atmosphere. In other words, if all over the planet wherever soils are warming because of climate change, you lose a quarter of your soil carbon, that's equivalent to doubling the amount of carbon dioxide in the atmosphere. So there's a huge potential for a feedback effect in which the warming from fossil fuel burning, which produces CO_2, causes ecosystems to release more CO_2, doubling the impact of the original fossil fuel burning. Ecosystems are playing a very scary role in the climate system. They act as a source of what we call positive feedback. And positive doesn't mean it's good for us: it means that it's reinforcing the warming by causing ecosystems to release carbon.

PE: And there are other side effects.

JH: To be sure. Very, very significant plant effects. The most dominant effect, the one that is probably the most important in the long run, is that if you look in the heated plots, a certain plant, a woody shrub called sagebrush—here's an example of it right here—grows sparsely in these meadows because we're at high elevation. But in the heated plots, the sagebrush is growing and dominating, outcompeting the other plants. So we're losing the forbs—the broad-leaved herbaceous flowering plants—and we're gaining woody shrubs. We anticipate that if this trend continues—and there's every reason to believe it will—that we would see that the heated plots will all be dominated by sagebrush. In the control plots, we have seen over the last twenty years a small increase in the amount of sagebrush. But in the heated plots, we're seeing a huge increase.

WILDFLOWERS VS. SAGEBRUSH

JH: Ranchers have used these meadows for grazing their cattle, and the cattle like to eat the forbs, not the sagebrush. So that's reason number one. The second explanation has to do with the local economy. This region, Gunnison County, is sometimes called the wildflower capital of the Rockies, and hundreds if not [more]—I think this year it will be at least a thousand people—flock to Crested Butte to take part in the wildflower festival. Will there be wildflowers here to see in forty years? We predict no. We predict that you'll have to have a sagebrush festival. But there's no reason to come up here from Oklahoma or Texas, you can stay back down there and look at the sagebrush.

MT: What about all the pollinators, Paul's butterflies?

JH: Exactly. Changes in the insect populations are largely driven by changes in the plant communities. So you have very different species of insects, fewer of the kinds that make a living pollinating plants because sagebrush is wind-pollinated. Some insects like to pollinate the yellow daisies. Others, the purple daisies. Some insects like to munch on the foliage of one kind of plant; other insects eat other kinds of plants. So as you shift plant communities, you cause shifts in the insect populations. Whether there will be an overall increase in numbers of insects or a decrease, we can't say yet. And if you ask me whether the world will be tolerable if we don't do anything about global warming, I'm really pessimistic. I think we're going to suffer enormous damage. Not just insect populations and entire ecosystem alterations. I'm talking about the general food supply and protection from the extremes of climate— drought and storms and fire. I would worry tremendously about all of the forest you're looking at here going up in smoke within the next forty years if we don't do something about global warming.

So I'm a pessimist about what the world will look like if we don't solve the problem. But I'm an optimist about the opportunities to solve the problem. I believe that the solutions are in hand. That they make economic sense and that we have the technology, the engineering, and the economics to back up clearly workable solutions to global warming—improving the efficiency of automobiles, replacing coal with solar—all of these things are in our hands; we can do them. What we lack is the political will to bring about these changes. And the reason I think it's so important isn't just because of global warming, but it really goes to the heart of our economy and our overall well-being. Nobody wants to see more oil spills in the Gulf of Mexico. Nobody wants to keep fighting wars in the Middle East over oil. And as long as we're dependent on foreign oil and burning coal, we're going to continue to have smog in the cities, mountaintop removal, acid rain, and global warming. But we could change all that.

THE FIRST STEP TOWARD CHANGE

JH: The single most important first step would be to get the money out of politics. As long as individuals can buy Congress, we're not going to see change. And as long as the coal and oil companies have this enormous power through the lobbying process and through campaign gifts, we're not going to see change, so we have to have campaign reform. Otherwise, the democratic process will not work.

And I'd like to mention something else in terms of this particular experiment.

MT: This heated meadow?

JH: Yes. This was the first experiment of its kind, to set up heaters, warm a meadow, and watch what happens. More than two dozen scientific publications have come out of this one experiment. It's been continuously funded by the National

Science Foundation for the past twenty years, and ten PhD dissertations have been carried out here. But the more important thing, in some ways, is that it was the first of its kind; it's the longest-running such experiment in the world. Today there are many others like it. But this was the first, and it continues. There are studies like this in bogs up in Minnesota, in forested land in Canada, in tundra in Alaska and Sweden, and we're going to be starting an experiment like this, if we can get the funding, in the tropics, in Puerto Rico. So this experiment has had a huge influence on the scientific community. It's initiated a whole agenda of field research in which we get away from just making model predictions of what ecosystems will look like, and we get really good firm data from the field. This year five different groups from other universities, not mine, have come here to make use of the site. Because when you have twenty years of warming, you have an opportunity to conduct all kinds of other studies that we wouldn't have thought of doing. Somebody is studying ant populations. Somebody else is looking at fungus. One of the students at University of California, Davis, is looking at evolutionary processes; because with twenty years of warming, some of the plants in these plots have had twenty generations now to adapt, and we're interested in whether there's been natural selection and plants are becoming adapted to the heating in the warm plots, so that evolutionary study is under way. In fact, the fellow behind me working in the meadow over there is gathering data as we speak.

MT: Do we even have the luxury of years left to conduct such research?

JH: That's important. Can we wait? Can we delay twenty years, thirty years before we take action? From just what we've learned in this experiment, yes, you might be able to wait five, ten, fifteen years. But the reason we can't wait is not because of what's going on in this sub-alpine meadow; the reason we can't wait is because each year the planet is being subjected to worse floods, worse droughts, worse fires, and a possibility that we're going past what we call tipping points where the climate won't recover from our fossil fuel burning. This is all real, and we don't want to wait for catastrophes and irreversible climate change. We should be trying to reduce emissions rapidly *now*. When I'm asked how fast should we act, I always say we should do as much as we can as fast as we can and as smartly as we can.

Ranchers should care because cattle don't like sagebrush. The economy of Crested Butte should care because the wildflowers here bring a huge amount of revenue to the county, as does snow in the winter ski season. But there's a much more important reason, even than those, and those are important enough.

THE FEEDBACK CRISIS

JH: The bigger reason is that these feedbacks that we've discovered—the loss of the carbon [in biological systems], creating more carbon dioxide in the air, and other feedbacks— these are telling us that global warming will be worse than we thought. Think about it this way. The models that predict that we're going to see, let's say, three degrees of warming—three degrees Celsius, remember, in the year 2040 or 2050—those models don't include ecological feedbacks. Those models were made by physicists, and they include a lot of important physics of the climate. But they don't include the ecosystem responses. When you include the ecosystem responses, as this experiment has taught us, you get more climate warming than you had bargained for originally. The problem is worse than we think, because ecosystems create positive feedbacks to the climate system, exacerbating the warming. So if the models now predict three degrees, we would say add another two or three to this. So that's really important. It means that when critics of global warming science say (invariably without any real knowledge of what they're talking about) that we're probably exaggerating the warming, the truth is we're *underestimating* the warming with our predictions because those predictions don't include the effects of ecological feedbacks. That's a really important message.

This is a microcosm. Most of the plants in this meadow are long-lived perennials. This northern long-stemmed fairy candelabra is one of my favorites. Look at this . . . see the aphids? Michael, you were asking about insects. Right there you can see the ants crawling up and down the stems. Essentially this makes the aphids competitors, and also the ants are going to tickle the aphids, and the aphids are going to poop, and the ants are going to eat the aphids' poop. Some people call it milking the aphid.

But in dry years, we've noticed more aphids in the control plots, and in wetter, very wet summers, we've noticed more aphids in the heated plot. And because snow melts earlier in the heated plots, the plants flower earlier. And so, for example, if we look at this heated plot right in front of us, you see a lot of yellow daisies growing near the heaters. And because there are more daisies in that plot under the heaters than under the control plots next to it, it looks like the heating must be causing more daisies to grow, but it's not. It is causing them to grow earlier. So we're looking at a particular time when the daisies are in bloom in the heated plots, but they haven't flowered yet in the control plots. When they flower in the control plots, about two weeks from now, there will be even more of them than there are in the heated plots. So we're changing the timing of the plants as well as the numbers of plants and numbers of flowers that they produce.

ECOLOGY OUT OF SYNCHRONIZATION

MT: So you have pollinators that are missing out on these flowers.

JH: Absolutely right. We call that the de-synchronization of the ecosystem. The ecosystem is literally being torn apart in time, because plants flower on a cycle that's governed by when snow melts here. But hummingbirds, which pollinate some of our plants, come here from Mexico, where they winter, and they arrive on a day-length cue, not the snowmelt cue. And so if plants are flowering earlier, and they flower before the pollinators get here, that's bad for the plants because they don't get pollinated, and it's bad for the hummingbirds and the other pollinators because they won't have a nectar source. It's a pure timing effect. These particular plants, these yellow daisies, of all the flowering forbs (of which there are about forty species), are probably the least affected by the heating in the sense that their numbers are still pretty sizable in the heated plots, as you can see. They are down compared to control plots, but not as much as the other flowering plants, and the reason is those are very deep-rooted, so they are able to tap into the deep soil moisture, whereas most of the plants here are more shallow-rooted, and the heaters, which dry the soil, deprive them of their soil moisture, and so they start to die back. And the other deep-rooted plant here is the sagebrush.

6 THE *EUPHYDRYAS* QUESTION

PE: One of the reasons I came to RMBL [Rocky Mountain Biological Laboratory] was to expand on the kind of research I had started at Stanford's Jasper Ridge. There are checkerspot butterflies here too, and I wanted to see if the things I was learning in the San Francisco Bay Area applied in the high mountains in Colorado. In one project, we worked at over 12,000-feet elevation at Cumberland Pass here in Gunnison County. It and many of the other high mountain areas around here have populations of a close relative of *Euphydryas editha*: *Euphydryas anicia*. But Cumberland Pass is not close by, and doing a mark-release-recapture experiment there is a pain in the ass, in part because at that elevation, running after them on steep rocky slopes can be tiring, but also because it takes about an hour and a half to drive out there. It violated one of the reasons that my wife, Anne, and I settled down here, because RMBL is a place where you can do field biology just stepping outside your door. But until recently you couldn't do it on checkerspots.

We've looked at *Euphydryas editha* butterflies all over the western United States, including at mid-range elevations in Gunnison County. But now we can work on another *Euphydryas* right here at the lab. This particular species, *Euphydryas*

Euphydryas gillettii. © P. R. Ehrlich

gillettii, is interesting for a number of reasons. It is relatively rare throughout its range. It looks like it belongs to the group of *Euphydryas* species that lives in Europe and Asia. Years ago we were doing research on *E. gillettii* around the Grand Tetons in Wyoming; they didn't occur as far south as Colorado. And one of the questions that we had to answer when we were studying them was what plants their caterpillars were eating. Another biologist had worked on this species in the northern Rockies, finding them feeding on twinberry honeysuckle (*Lonicera involucrata*).

We quickly confirmed that that was what the populations we were looking at laid egg masses on and the caterpillars fed on, and it is a plant common at RMBL. In fact, the adults in the Tetons were taking nectar from flowers that are also abundant around the lab here in Gothic and one that is also used by local *Euphydryas*. So the immediate question arose:

Why wasn't *Euphydryas gillettii* at RMBL? It was a question in the area of study called "community assembly"—trying to understand the mix of kinds of plants and animals found in a given place. Was it that the butterflies gradually moving into North America were unable to get across the Red Desert gap in the Rockies in southern Wyoming? Or had they penetrated to the Gothic area but been unable to survive? Was the habitat really suitable, or was a climate factor or predator we hadn't recognized barring *E. gillettii* from the lab area?

With student Cheri Holdren, I wanted to find out, since the food plants and the nectar sources here are ideal. So in the summer of 1977, we transplanted some of the butterflies here. We moved a lot of individual egg masses from Wyoming (the transplant was made easy by our being able to transport them in a light aircraft). But they didn't seem to take very well. We thought that I had made a stupid mistake by putting them out on a north-facing slope, where snow often persisted very late in spring.

So we decided we'd do another transplant and put the immigrants down at Cement Creek, a lower-altitude site near here with the requisite plants, and place them on a roughly south-facing slope. And we did that, and the butterflies did exactly the opposite of what we expected. Namely, where we thought they would go extinct at the lab, they hung on; and where we expected them to hang on, they went extinct. As far as we know—I haven't checked in the last few years—there's never been any sign of them after the first year at Cement Creek. But they hung on by their wingtips across the valley here, and for decades we used to go and check periodically.

BIO-INVASIVES

PE: One of the key open questions the *E. gillettii* research applies to concerns the general rules for when a species will be-

come invasive. Consider the house (or "English") sparrow—
Passer domesticus. Transplant it anywhere, it establishes a colony
and usually spreads widely and takes off. I can remember driving all night long through a stony desert in South Africa to get
to a bird-banding station way the hell out in the desert. And I
can remember getting up at five the next morning, grabbing
my binoculars, and going out to add to my life list as quickly
as I could. And there was a bird, hopping across the stony part
of the farmyard—a house sparrow!

In about 1870, I think, some British biologists transferred
the English tree sparrow, a close relative of the house sparrow
(*Passer montanus*), to St. Louis. There it has stayed. If you go
to St. Louis today, you can find them. It occurs in about four
states, but it's where four states come together, and there's this
tiny little range, and that's it. Where the two have been introduced elsewhere, results have often been similar. And nobody
knows for sure what the difference is between *Passer montanus*
and *Passer domesticus*. What makes one of them a brilliant invader and the other one not-so-hot?

So, with respect to butterflies, one of the questions we
were trying to answer is: What would happen if we turned
this butterfly into an invader? And for quite a while we could
find a few individuals, but never up the Copper Creek Trail,
where I spend much of my time, or out on Trail 401, which I
also ritually patrol. I've never seen one in butterfly-rich Cross-
ing Meadow, about a mile up the creek from here. But then,
surprise, surprise: the *E. gillettii* population exploded in 2002,
a quarter century after they were introduced. Cheri is too busy
being involved in environmental politics (she's married to my
close friend and colleague John Holdren, who is President
Obama's science adviser), but my wonderful colleague Carol
Boggs and her students have taken the lead in the study.

Every day now, when the *E. gillettii* are flying, I go out and check along 401, look down the road, look along the trail going to the main road, and so on, to look for signs that the population is expanding. And the population has fluctuated.

Carol's got one or two PhD students trying to figure out what's going on, such as who is eating the *E. gillettii* and who isn't eating them.

The project shows two aspects of field biology. One is that even after thirty-five years of study, you still may not have a clear answer to your question. The other concerns the ethics of field manipulations. Was it ethical to possibly create a species invasion with unknown consequences for the local biotic community? I think so, based on what I knew about both the organism and the community—and the seriousness of the invasive species problem for global biodiversity. Colleagues agreed; I got permission from RMBL to do it. But, of course, others may differ!

When I was a kid, my biggest worry was that we would know everything in science before I got a chance to become a scientist. It was the silliest worry I ever had in my entire life! We know an enormous amount about these butterflies in terms of their population biology—more about butterflies than any other invertebrates, probably, except possibly ants and bees. And I'm pretty sure that we know more about the plants that butterflies eat than we do for any other large group of herbivorous insects. Nonetheless, why we originally didn't find *E. gillettii* around RMBL remains a mystery.

ISSUES FACING HUMANITY

PE: I find butterflies fascinating in themselves. But I find that the issues of sustainability facing humanity are more than interesting. Now I've shifted my attention from theoretical ecological-evolutionary questions that might be answered by a butterfly system and am focused on trying to figure out how humanity can keep a world together that will support human beings in the long term and will give them time to evolve ways to treat each other kindly. The manipulations that we do in an experiment like that on *E. gillettii* are really trivial compared to what's going on all over the globe right now, where thousands of populations are being forced to extinction daily, and where people are adding carbon dioxide to the atmosphere ad lib with only scattered and inadequate attempts to stop, even though almost every sensible scientist knows it could be lethal for humanity. In a sense *Homo sapiens* is running a vast experiment on itself. Whether you move a rare butterfly a few hundred miles or not is trivial when trying to understand the world, as opposed to the experiment we're doing collectively to see if we can destroy our life-support systems. So my conscience is clear on this one.

We have, after all, just had maybe the biggest geological event on this planet in terms of biodiversity in something like the last five or ten million years. It elevated the level of Central America so that the Pacific and Atlantic were divided, with huge biotic consequences. Now the North Atlantic and the North Pacific have been reconnected through the Arctic. The flows are now moving, and they are picking up Pacific organisms in the North Atlantic. We don't have any idea what the effects are going to be, except they could easily be very traumatic on creatures like certain fish or perhaps on oysters, which are disappearing over much of North America. The scale of all this, in terms of bio-invasives, could be huge. Humanity hasn't created a total agricultural disaster—not yet; climate disruption probably hasn't started eating through the wheat belt or the corn belt. That's why John Harte's work is so important, illuminating the potential effects of climate disruption.

Most of the things our group studies now are studied not for their own sake, but for what they can tell us about what we can do for each other, for humanity, and for the rest of biodiversity. For example, our group has spent a lot of time and effort, under Gretchen Daily's leadership, looking at how critical ecosystem services to agriculture can be preserved in tropical landscapes.

7 THE LESSON OF ACORNS

CHANGE IS UBIQUITOUS

PE: Because of our own life spans, we tend to think that the world is absolutely permanent and without dramatic climate disruption or other big-time change. But, of course, change is always going on at a pace that's rapid in ecological time, but not on the time scale we personally perceive. We tend to perceive our ecological background as a constant. On the Stanford campus, there's a hiking area called Dish Hill (it features a huge radio telescope dish). It has large stands of live oak trees that look permanent. But you have to look very closely. Is there recruitment in the groves? The answer is that most acorns are now eaten by abundant ground squirrels, acorn woodpeckers, and others. There are few seedlings. When the adult oaks die, that will be it for the groves of trees.

If you go to the tropics, you can see something similar. Very often when they clear a rain forest, farmers will leave a rain forest tree sitting in the middle of their pasture so that the cows, or whatever they are raising, can have shade. Rain forest trees can live in a pasture for a hundred years. It has been shown very clearly that they are extremely valuable for maintaining bird populations. Birds can often survive in mixed agricultural countrysides, in part because of the presence of these specimen rain forest trees. But of course, they're zombies; there is no reproduction. They are going to be gone soon. So you don't want to assume—as so many biologists used to fifty, a hundred years ago—that what you're seeing is a permanent vegetation pattern. The world is in constant flux.

One dramatic demonstration of the constant flux was produced by geographer Garry Rogers, who collected scenic nineteenth-century photographs from the Great Basin. Then he located the places the shots were taken and took new ones a century later. The changes in the plant formations were dramatic.

Now, of course, we're facing the biggest climatic change in human history, and likely the biggest overall changes in environment and social structure civilization has ever experienced. Everything changes all the time, but humanity tends to be increasing the *rates* of change. Don't forget that, and be ready for an even greater increase in the rate of transformation.

MT: Well, we now know from recent data compiled by researchers from the University of Milan, for example, that the glaciers in Solu-Khumbu, at the base of Mount Everest, have receded some 13 percent in the last fifty years. The bottom level of the snow line is evidently fifty meters higher than it was a half century ago. That's striking. For ten years, I photographed a glacier in Ladakh from the exact same spot, year after year. The glacier retreated well over a mile during that decade. I have seen this happening firsthand, and I don't give any credence to the conservative right wing that has its own agenda and argues stridently that for over eight hundred years the planet has witnessed "Little Ice Ages" in an effort to combat the literally tens of thousands of peer-reviewed scientific

Mount St. Elias, Northwest Face, Alaska. © M. C. Tobias

papers that overwhelmingly draw a draconian and all-too-real picture of climate change, which, finally, the Obama administration and most governments worldwide now recognize to be the new ecological normal.[1]

Say, it's getting dark. A lot of mosquitoes. Time to go, don't you think? West Nile is spreading throughout the western United States, speaking of climate change.

PE: In fact, it's later than you think.

---

1. Jason Burke, "Mount Everest's Glaciers Shrinking at Increasing Rate, Say Researchers," *Guardian*, May 23, 2013, http://www.guardian.co.uk/world/2013/may/23/mount-everest-glaciers -shrinking-global-warming.

AFTERWORD

A mere year after Paul, Michael, and John had their conversation, things looked even more dire than they had thought. One of the worst droughts in history had parched much of the United States, and many other "unusual" climate-related events were occurring at record-breaking frequencies. The corn and soybean crops suffered greatly (at least one farmer was reported to be feeding candy corn to cows); the Indian monsoon suffered a partial failure; floods ravaged parts of Asia; Hurricane Sandy wreaked havoc on much of the coastal areas of the eastern United States; and Australia baked under extreme record heat. The climate instability predicted by the models was building. And virtually daily there were reports of likely climate-induced problems with the human food supply. There are increasing signs that the global climate has been bounced out of the relatively benign and constant mode within which agriculture and civilization developed over the past ten centuries. It looks like all bets on the future are off.

As mentioned earlier in terms of Mount Everest, a one-day alpine-style ascent of Everest is easier now—fewer crevasses with which one need contend. Little consolation, though, for all the species—humans included—depending on the annual snowmelt in the sub-Gangetic, Indian subcontinent.

Locally, the famous flowers of the Gothic area never appeared in 2012—the density of blooms was perhaps one-hundredth of normal, and butterflies jostled one another trying to suck nectar from the few remaining flowers. There was severe fire danger as the vegetation dried, and the usual snow patches disappeared from the mountains. Such events are predicted to be statistically more common—just as next year Gothic may be smothered in snow. As the planet warms, more water can be held in the atmosphere, leading to heavier rains when it is above freezing and heavier snows when below. Contrary to beliefs of deniers of every persuasion, heavier snows can be expected with warming wherever it remains possible for temperatures below-zero Celsius to persist. And John Harte has not only found dramatic and scary changes in his heated plots, but his unheated plots are following slowly his heated ones. Nature, in the guise of real global warming, is copying his experiment.

The 2012 summer did bring one big surprise in Paul's experiment. Paul's colleague Dr. Christine Turnbull got a picture of a *Euphydryas gillettii* just beyond the end of the transect Paul had searched each summer looking for any signs of the butterflies' spread from the introduction site. They went to the spot, but there were no food plants, and Paul guessed that Chris had just been lucky enough to see a stray. He continued farther along the trail and just before turning back spotted some *Lonicera*. Even before reaching the plants, he could see *E. gillettii* egg masses on their leaves. It turned out there were hundreds of egg masses there and beyond. The largest population of the transplanted butterfly had apparently been established years before, likely in 2002, just beyond the original search radius! So the mystery intensifies, and 2014 should be an exciting year. If this all sounds somewhat confusing, there is one unambiguous message to glean from it: Field biology is always full of surprises!

Biologists are accustomed to change and surprise. But in August 2012, Republicans in Congress announced that they didn't believe in global temperature increase. No surprise, but the authors guarantee that in coming years as water scarcity affects everyone, such adamant skepticism will vanish. Remember the eerie silence that descended upon all those who, in the 1930s, had insisted the dust bowl had nothing to do with man-made causes, like the rapacious exhaustion of fragile soils for cash crops.

Nonetheless, the eleventh-hour fumbling on the edge of the so-called "fiscal cliff" at the start of 2013 demonstrated conclusively that the human species prefers to play chicken in the face of deficits—financial deficits or ecological deficits. Indeed, this practice of fumbling may be a general result of the interactions of genetic and cultural evolution, given the all-too-clear history of human denial in the face of ecological illiteracy, where the outcome has spelled disaster for one human civilization after another. Our brash indifference to the consequences of our actions has spelled near doom, from the collapse of society on Easter Island to Three Mile Island. But because our finite Earth is also an island, and *Homo sapiens* is quite literally the stuff of that island Earth, it is now only too apparent that the current 7.2 billion members of this human club must stop fumbling on cliffs and, rather, find ways to live peacefully and sustainably.

To that end, the authors propose the following few personal steps that would, if fully embraced by the majority of those alive in this human generation, go a long way toward ensuring a future for at least some viable portion of our species, and for populations of other species well into the future.

1. Have no more than one child, or none at all.
2. Try to reduce your consumption, one item and one day at a time.

3. Try to retrain your taste buds—it is possible—so as to eat fewer animals and more plants. Walk or ride a bicycle whenever you can, rather than drive.

4. Ignore those who would ridicule some of the old clichés about living as nothing other than stupidity. Indeed, many of them are as inspired today as when they were first enunciated. Such as: Make love, not war (just use protection when you do); Follow the Golden Rule (whichever one you prefer, but don't measure it in troy ounces); Try not to kill, whether a dolphin, a bird, or a butterfly.

5. Plant a native tree, a native shrub, anything that is native and will enrich the soil and the quality of life for others, most notably pollinators: birds, bees, wasps, butterflies, bats, and other animals.

6. Become environmentally informed, preferably *before* casting your next vote or engaging in your next purchase.

7. Remember we are social animals and really enjoy working with others. Join an environmental group or a group striving for peace or social justice. There are thousands of important NGOs. Here are a few of them: the Millennium Alliance for Humanity and the Biosphere (MAHB, making humanity "future smart"; mahb.stanford.edu), 350.org (to solve the climate crisis; http://www.350.org/), the Sierra Club (working for general environmental sanity; http://www.sierraclub.org/), and Dancing Star Foundation (working worldwide to sensitize people to conservation and to animal rights; www.dancingstarfoundation .org).

8. Above all, *do something*. Don't face a deteriorating future with equanimity. Remember the old navy saying: "If you can keep your head while all those around you are losing theirs—you simply don't understand the situation!"

And to end on a happy note, the scientific community has begun to shout out about the human predicament. A recent example is the Consensus Statement of Global Scientists (reprinted in the appendix). We, too, add their shout here—let's hope the public and the politicians are listening. They are in China at least, as demonstrated by Michael's article in recommended reading.

APPENDIX: ESSENTIAL POINTS FOR POLICY MAKERS

SCIENTISTS' CONSENSUS ON MAINTAINING HUMANITY'S LIFE SUPPORT SYSTEMS IN THE TWENTY-FIRST CENTURY

Earth is rapidly approaching a tipping point. Human impacts are causing alarming levels of harm to our planet. As scientists who study the interaction of people with the rest of the biosphere using a wide range of approaches, we agree that the evidence that humans are damaging their ecological life-support systems is overwhelming.

We further agree that, based on the best scientific information available, human quality of life will suffer substantial degradation by the year 2050 if we continue on our current path.

Science unequivocally demonstrates the human impacts of key concern:

- *Climate disruption:* There has been more, faster climate change than since humans first became a species.
- *Extinctions:* Not since the dinosaurs went extinct have so many species and populations died out so fast, both on land and in the oceans.
- *Wholesale loss of diverse ecosystems:* We have plowed, paved, or otherwise transformed more than 40 percent of Earth's

ice-free land, and no place on land or in the sea is free of our direct or indirect influences.

- *Pollution:* Environmental contaminants in the air, water, and land are at record levels and increasing, seriously harming people and wildlife in unforeseen ways.
- *Human population growth and consumption patterns:* Seven billion people alive today will likely grow to 9.5 billion by 2050, and the pressures of heavy material consumption among the middle class and wealthy may well intensify.

By the time today's children reach middle age, it is extremely likely that Earth's life-support systems, critical for human prosperity and existence, will be irretrievably damaged by the magnitude, global extent, and combination of these human-caused environmental stressors, unless we take concrete, immediate actions to ensure a sustainable, high-quality future.

As members of the scientific community actively involved in assessing the biological and societal impacts of global change, we are sounding this alarm to the world. For humanity's continued health and prosperity, we all—individuals, businesses, political leaders, religious leaders, scientists, and people in every walk of life—must work hard to solve these five global problems, starting today:

1. *Climate Disruption*
2. *Extinctions*
3. *Loss of Ecosystem Diversity*
4. *Pollution*
5. *Human Population Growth and Resource Consumption.*

The full statement (http://mahb.stanford.edu/consensus -statement-from-global-scientists/) has been signed by 520 global scientists from forty-four countries. Those signatures were obtained within a month of completion of the statement, by direct e-mail requests from the authors and their close colleagues to a targeted group of well-regarded global change scientists. The signers include two Nobel Laureates, thirty-three members of the U.S. National Academy of Sciences, forty-two members of the American Academy of Arts and Sciences, and several members of various European scientific academies.

.

SUGGESTED RESOURCES

DR. EHRLICH'S SUGGESTIONS

Diamond, J. M. *Guns, Germs, and Steel: The Fates of Human Societies.* New York: Norton, 1997. *The classic anthropological work on geographic differences among human societies.*

————. *The World until Yesterday.* New York: Viking, 2012. *Discusses many things we might learn from the ways our ancestors lived in traditional societies. As interesting as it is instructive.*

Ehrlich, P. R. "Ecoethics: Now Central to All Ethics." *Journal of Bioethical Inquiry* 6 (2009): 417–36. *One ecologist's views on ethics.*

Ehrlich, P. R., and A. H. Ehrlich. "Can a Collapse of Global Civilization Be Avoided?" *Proceedings of the Royal Society B* 280 (2013). *Short and heavily documented paper, if you wish to look up scientific articles dealing with many of the issues we discuss in this book.*

————. *The Dominant Animal: Human Evolution and the Environment.* 2nd ed. Washington, DC: Island Press, 2009. *Overview of the human predicament.*

Ehrlich, P. R., and R. E. Ornstein. *Humanity on a Tightrope: Thoughts on Empathy, Family, and Big Changes for a Viable Future.* New York: Rowman & Littlefield, 2010. *A discussion of the ways civilization might move to avoid its own collapse.*

Fauchere, Christophe, dir. *Mother: Caring for 7 Billion.* 2011. http://www.motherthefilm.com/. *A fine film dealing with the population problem.*

Gardner, Dave, dir. *Growthbusters: Hooked on Growth.* 2012. http://www.growthbusters.org/. *Maybe the most important movie ever made, tackling the nonsense of perpetual growth.*

Harte, J., and M. E. Harte. *Cool the Earth, Save the Economy: Solving the Climate Crisis Is Easy*. 2008. http://cooltheearth.us/. *Free-to-download excellent book.*

Huesemann, M., and J. Huesemann. *Techno-Fix: Why Technology Won't Save Us or the Environment*. Gabriola Island, BC: New Society Publishers, 2012. *Good cure for those who think technology can breach the laws of nature.*

Klare, M. T. *The Race for What's Left: The Global Scramble for the World's Last Resources*. New York: Metropolitan Books, 2012. *Best recent analysis of humanity's resource dilemmas.*

Oreskes, N., and E. M. Conway. *Merchants of Doubt: How a Handful of Scientists Obscured the Truth on Issues from Tobacco Smoke to Global Warming*. New York: Bloomsbury Press, 2010. *Great historical look at how the few in "Murder Incorporated" (think the tobacco, fossil fuel, and firearms industries) can confuse the many for profit and ignore the many millions of deaths for which they have been and will be responsible.*

DR. TOBIAS'S SUGGESTIONS

Buttner, Nils. *Landscape Painting: A History*, translated by Russell Stockman. Abbeville Press, 2006. *By far the most extensive overview of ecological aesthetics ever written.*

Clough, G. Wayne. *Increasing Scientific Literacy: A Shared Responsibility*. Washington, DC: Smithsonian Institution, 2011. http://www.si.edu/Content/Pdf/About/Secretary/Increasing-Scientific-Literacy-a-Shared-Responsibility.pdf.

Glacken, Clarence. *Traces on the Rhodian Shore: Nature and Culture in Western Thought from Ancient Times to the End of the Eighteenth Century*. Berkeley: University of California Press, 1976. *Glacken's masterpiece remains one of the most evocative and deeply woven history of ecological thought yet written.*

The Jaina Sutras, translated by Hermann Jacobi. In *Sacred Books of the East*, vol. 22, edited by Max Muller. 1884. *The Jain beliefs, as outlined in the* Jaina Sutras, *comprise some of the most profound recommendations for human restraint and non-violence ever advocated.*

Matthiessen, Peter. *Wildlife in America.* New York: Viking Press, 1959. *Matthiessen's chronicle of extinctions ranks as one of the most depressing and telling books of its kind.*

Mittermeier, Russell A., Patricio Robles Gil, Michael Hoffman, John Pilgrim, Thomas Brooks, Cristina Goettsch Mittermeier, John Lamoreux, and Gustavo A. B. da Fonseca. *Hotspots Revisited.* Chicago: Conservation International/Cemex/University of Chicago Press, 2003.

Mowat, Farley. *Never Cry Wolf.* New York: McClelland and Stewart, 1963. *The film by the same title, based upon the book, was directed by Carroll Ballard and released by Walt Disney Pictures in 1983.*

Ovid (Publius Ovidius Naso). *The Metamorphoses. The great Latin narrative poem, completed in* AD 8, *forms the basis for one of the most influential dramas of ecological intrigue, text, and subtext ever invented. In countless translations in nearly every language.*

Sabin, Paul. *The Bet: Paul Ehrlich, Julian Simon, and Our Gamble over Earth's Future.* New Haven, CT: Yale University Press, 2013.

Schama, Simon. *Landscape and Memory.* New York: Vintage, 1996. *One of the world's foremost literary critics and historians points his acutely sensitive gaze upon the interdisciplinary enigmas of the impact of nature on the human psyche and human history.*

Schell, Jonathan. *The Fate of the Earth.* New York: Knopf, 1982. *First anthologized in three parts in the* New Yorker *magazine, Schell's classic work chronicles the perils of nuclear war that remain all too with us.*

Thoreau, Henry David. *The Writings of Henry David Thoreau*, http://thoreau.library.ucsb.edu/writings_journals.html. *Thoreau's multiple notebooks provide one of the great philosophical and continuous records of nature ever written.*

Tobias, Michael Charles. *The Adventures of Mr. Marigold*. First published in New Zealand in 2005, now available from Michael Charles Tobias Books on Amazon. *The one book of mine (an 1,836-page reverie of ecological contradictions, a novel) that I can safely recommend as indicative of the Utopian world I should like such dialogues, as that engaged in herewith, to at least, in part, intimate.*

———. "China Declares Global State of Emergency: An Urgent Telegram from Taihu." *Forbes*, May 21, 2013. http://www.forbes.com/sites/michaeltobias/2013/05/21/china-declares-global-state-of-emergency-an-urgent-telegram-from-taihu/.

———. *World War III: Population and the Biosphere at the End of the Millennium*. Preface by Jane Goodall. New York: Continuum Books, 1998. *The second, revised edition of the author's research in nation after nation examining the human population explosion and resulting biological fallout.*

Tobias, Michael Charles, and Jane Gray Morrison. *Sanctuary: Global Oases of Innocence*. Preface by the Queen of Bhutan, Her Majesty Ashi Dorji Wangmo Wangchuck. Los Angeles: Council Oak Books/Dancing Star Foundation, 2008. *A celebration of success stories in twenty-four countries from the front lines of conservation biology and animal rights.*